Floyd Clymer's
MOTORBOOK
Special

Editor's Note:

Putting together a book like *The Best of HOT ROD* was both a pleasure and a source of frustration. The pleasure was reviewing all the great issues of HOT ROD. The frustration was knowing that there weren't enough pages to include all the great photos. It was also frustrating to realize how lightly we took ourselves in the early years. Until 1955 there was no formal filing system for photos, manuscripts and back issues. Many files were discarded, lost during numerous moves or even damaged when storage rooms flooded. As a result it is possible that a favorite photo of yours may not have made the book.

A lot of terrific photos did make it, though. Some cars weren't featured in HOT ROD even though they were well-known cars. We used photos that were run in HOT ROD only, although in some cases an unpublished picture was used for variety.

The photos are identified by year and month so that readers desiring more information on a subject can look it up in the magazine. Some cars appeared more than once, and in those cases the date refers to the major story on the vehicle.

We hope you enjoy this book because it would be our pleasure to reopen the files for further installments of *The Best of HOT ROD*.

This Book is Gratefully Dedicated
to the Founder of HOT ROD,

ROBERT E. PETERSEN.

The Best of HOT ROD

Contents

PETERSEN AUTOMOTIVE BOOKS

LEE KELLEY/Editorial Director
BRUCE CALDWELL/Editor
CRAIG CALDWELL/Editor
JACKIE ANDERSEN/Managing Editor
CHARLIE HAYWARD/Art Director
LINDSAY BOLYARD/Art Production
LINNEA HUNT-STEWART/Copy Editor
LINDA SARGENT/Copy Editor
ANNE SLATER/Copy Editor
FERN CASON/Editorial Coordinator

DICK VAN CLEVE/Publisher, HOT ROD Magazine
DICK DAY/Senior Vice President, HOT ROD Publications Division

When the early photo credits read, "Photo by Pete," it meant Robert E. Petersen was out there getting the latest event coverage for HOT ROD. Below, he poses with 240 issues of HOT ROD Magazine on the occasion of HRM's 20th anniversary.

Cover: The nostalgic garage scene was staged and photographed by Elliot Gilbert in collaboration with Art Director Charlie Hayward. Models Loren Cedar, Dave Cameron and Gary Sexton added just the right touch to Ed Newett's fine '32 Ford roadster. Jim and Patti Jacobs provided the nifty garage, and Pete Eastwood supplied the free advice.

Introduction

HOT ROD Magazine—they said it would never get off the ground, and probably with good reason. "Hot rodders" were considered outlaws by the automotive establishment, and to the general public we were just Peck's Bad Boys. But no one, including ourselves, properly estimated the number of devoted rodders in the early years after World War II or the impact this group of enthusiasts would have on the automotive industry in the years to follow.

Work on the first issue of HOT ROD actually began in 1947. I had built cars and played around with them, but at that time I was working in the publicity department of MGM Studios. Bob Lindsay—my friend and HRM cofounder—and I saw what was happening on the streets and at the dry lakes. With the help of Lindsay's father, who had a magazine, we started on the first issue. About that same time I left the movie studio to start a company called Hollywood Publicity Associates. One of our clients was Earl "Mad Man" Muntz, and we were working on a plan to develop Southern California's first real dragstrip. The dragstrip was never built, but one of the fund raising ideas, putting on a hot rod car show, helped seed publication of the first issue. That first show was held in the Los Angeles Armory in January 1948 and was a financial and public relations success. The general public got a chance to view the quality and engineering of the rods and customs, while Hollywood Associates and show co-sponsors Southern California Timing Association got some much needed working capital. Being hawked outside the doors of the Armory and at a booth inside, was Volume One, Number One of HOT ROD Magazine.

The first print order of 10,000 copies was a sellout, but in those early years, Petersen Publishing was no major economic threat to the country's publishing houses. We had a one-room office, a typewriter and a lot of imagination. The phone came sometime around the third issue of HRM. Those first issues were basically written on stone. That is, we'd take some refreshments, go to the printer and work all night putting the issue together. We'd often have to get on the phone and ask guys things like, "What's the name of the guy in the Deuce highboy who wears the stocking cap?" But somehow we got those issues out. Circulation was a little different in the early days from what it is now, also. We'd take our issues and a change purse and hit the races, car shows and speed shops. At night we'd go to the drive-ins, sell a couple of subscriptions and then go buy dinner with the sub money.

We became more sophisticated and found out about wholesale distributors. As soon as an issue was printed, we'd load bundles of magazines into the company car, a '29 Model A Roadster pickup, and deliver them to the distributor. As sales grew, HOT ROD attracted more advertising, and the number of pages and size of the staff grew too. Wally Parks came over from SCTA to become HOT ROD's first real editor. There were a lot of other good guys who helped out in the early days, guys like Ak Miller, Eric Rickman, Bud Coons, Lee Ryan, Robert Gottlieb and Bill Burke. They really helped us, and they're all still involved in the industry today.

From those uncertain beginnings HOT ROD has grown into the world's largest selling automotive magazine. The term "hot rodder" is no longer a dirty word, and today's "hot rods" are now worth more than most foreign exotic cars. Getting from there to here was not easy, but it was always fun. Many, many excellent people have worked on the staff under editors Wally Parks, Bob Greene, Ray Brock, Jim McFarland, Don Evans, A. B. Shuman, Terry Cook, Jim McCraw, John Dianna, Lee Kelley and Leonard Emanuelson.

This book, while highlighting many of the truly great moments in hot rodding, is not intended to be a formal history of the sport or of HOT ROD Magazine. Our intention is to look back at some of the great cars, great people and great events that have been documented in nearly 400 issues. Without a doubt we have probably left out some of your favorite photos. Some, especially from the earliest years, have been lost or damaged. Others, there just wasn't room for. But while we have always believed that the best is yet to come, the Best of HOT ROD to date is pretty darn good.

Robert Petersen

Flat Out on the Flats

Hot rodding didn't start with HOT ROD Magazine and neither did the concept of going flat out on the flats. But in the early days of post-war hot rodding, the flats were where it was at and that's where you'd find the staff of HRM. Muroc was the finest of the dry lakes. It was close (90 miles northeast of Los Angeles), and it was a way of life for many So-Cal racers to drive up Friday night, camp out next to their roadsters, and run what they brung (after removing windshields and headlights) the next day.

Then came El Mirage. Speeds got

November 1952 The HRM staff would often go to great lengths to get the story or the photos. Manuel Ayulo's '27 track roadster was converted to the HRM "camera car" for the 1952 Bonneville speed trails.

faster. More and more cars were showing up each weekend and the Southern California Timing Association, formed back in 1937, began the battle to make the lakes safe from the accidents and public outcry against hot rodders that threatened to shut down the beloved flats. Working hand in hand with the SCTA was the young staff of the fledgling HRM.

It wasn't long before another crisis threatened to put an end to running on the flats. The dry lakes were literally breaking up, their relatively fragile surfaces showing the effects of years of constant use. Accidents were again on the rise on the rutted courses of El Mirage, Rosamond and Harper. So in 1949 a trio of hot rod-oriented representatives met with officials of the state of Utah to discuss opening the Bonneville salt flats to a yearly speed week. Returning to Los Angeles with the provisional okay were HRM Publisher Robert Petersen, Petersen Publishing Co.'s General Manager Lee Ryan and SCTA Executive Secretary (and soon to be HRM Editor) Wally Parks. HRM co-sponsored the first Bonneville Nationals that year and the rest, as they say, is history.

In the 1950s, Bonneville was a major part of HRM editorial content. El Mirage and Rusetta were still covered from time to time, but HRM's Bonneville involvement, from sponsoring Alex Xydias' So-Cal Special streamliner to staffers' personal lakes cars (including an endless string of innovative designs from Advertising Manager Bill Burke) to the previews of cars being built in the backyards of hot rodders with names like Mickey Thompson and Ak Miller, was everywhere.

As HRM and hot rodding grew to include activities such as drag racing, customs, street machines, boating and the like, the pages devoted to Bonneville gradually decreased but never died. The quest to break 400 mph, then 500, 600 and, eventually, the speed of sound barriers kept the interest in the flats from going flat.

And still today, as in '49, for one week each year the editorial offices empty out as the HRM staff makes its annual trek to the salts of Utah. As long as there are dry lakes and hot rodders, HRM will continue to be active members of the Flats Earth Society.

March 1948 The track roadster belonged to Don Blair, the grinning face to HRM's own Stroker McGurk, Tom Medley.

April 1950 Bob and Dick Pierson's cover coupe was based on a '34 Ford body, but with its radical chop and track-style nose it had cleaner lines than most roadsters. Power for its 142-mph record run came from a 59-A flathead bored and stroked to a 276-cubic-inch displacement. The Edelbrock manifold carried three Stromberg 48's converted for use with alcohol. For their record runs they used 2.94 gears in the Halibrand quick-change rearend. Tires were 7.50x18 Firestones. (Photo: Tom Medley)

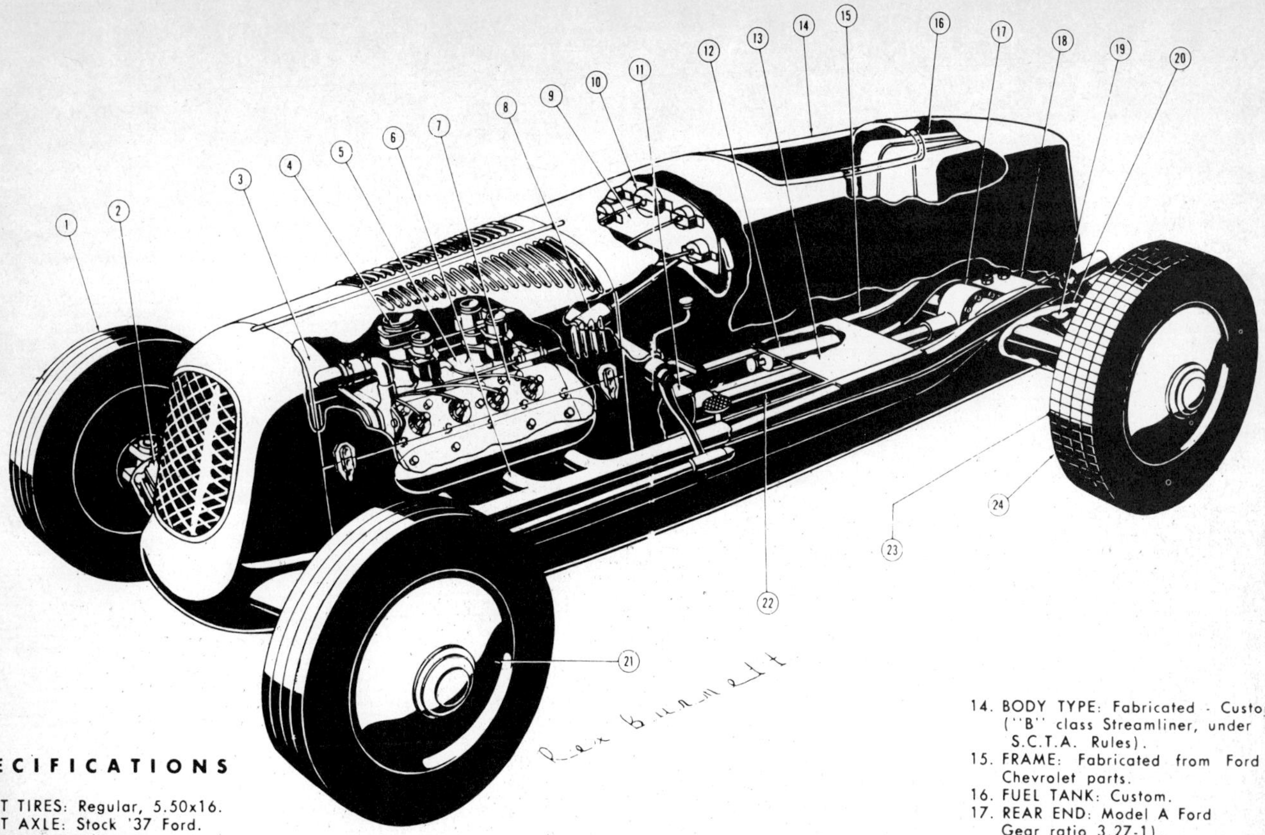

Rex Burnett

SPECIFICATIONS

1. FRONT TIRES: Regular, 5.50x16.
2. FRONT AXLE: Stock '37 Ford.
   FRONT SUSPENSION: None
   (Solid Axle, bolted to frame)
3. RADIATOR: Custom Fabricated.
   (4" core, tanks on top and bottom)
4. CARBURETORS: Stromberg, 3 singles,
   1 double-dual throat, for equal distribution.
5. MANIFOLD (exhaust and intake): Miller.
   IGNITION: Scintilla Vertex Magneto
   PISTON DISPLACEMENT: 239 cu. in.
   GERRY GRANT'S PISTON RINGS
   CAM: Miller
6. HEADERS: Miller.
7. ENGINE: '34 Ford
   Bored to 3-3/16 Standard
   (.125 over Ford)
   Stroke: Standard 3¾".
   HEADS: Stock Ford (Filled and Re-chambered; 9-1 estimated ratio).
8. STEERING: Franklin.
9. DASH BOARD: Custom
   (5 instruments)
10. TACHOMETER: Stewart-Warner.
11. TRANSMISSION: Ford, but has custom overdrive in place of 2nd gear. (For standing starts uses Standard 2nd gear)
    GEAR CHANGES: Only for overdrive in place of 2nd gear.
12. FUEL PUMP: Offenhauser (with Hilborn Injector System set-up)
    Regular Hand Pump (with 4 carburetor system set-up).
13. SEAT: Sheet Steel Cross Member built up.
14. BODY TYPE: Fabricated - Custom ("B" class Streamliner, under S.C.T.A. Rules).
15. FRAME: Fabricated from Ford and Chevrolet parts.
16. FUEL TANK: Custom.
17. REAR END: Model A Ford
    Gear ratio 3.27-1).
18. CROSS MEMBER: Fabricated sheet steel and tubing.
19. REAR SUSPENSION: '32 Ford
    (Turning axle housing upside down and reversing them).
20. SHOCKS: Model A Ford
    (Central and Back Tail cone).
21. WHEELS: Ford and customed.
22. DRIVE SHAFT: Ford - cut to length.
23. BRAKES: '40 Ford Hydraulic
    (rear only).
24. REAR TIRES: Indianapolis Speedway 5.50x17.

April 1948 The subject of Rex Burnett's cutaway and Hot Rod of the Month for April was "old-timer at the lakes" Stuart Hilborn's jet black B class streamliner. In a classic understatement of things to come, the article stated "Hilborn sometimes uses a fuel injection system which he has designed and built." (Drawing: Rex Burnett)

May 1950 Bill Kenz brought his July 1949 HRM cover truck from Denver to run a "surprising" 140 mph during Bonneville Speed Week 1949. The twin flathead-powered '31 Model A pickup, named the "Odd Rod of the Month" in its initial HRM appearance, was featured in the May '50 Bonneville preview called "Safer on the Salt."
(Photo: Eric Rickman)

November 1950 The So-Cal Special was called America's most outstanding hot rod by the magazine that coincidentally sponsored it. Built by Alex Xydias and Dean Batchelor, the car set top speed honors in both '49 and '50. The '49 Mercury V8 Sixty engine sat between modified T frame rails and was wrapped in a Valley Custom-built aluminum body. (Photo: Maynard & Harris)

October 1950 The Bonneville Nationals were the big event in the early years and the HRM staff turned out in force to bring back exclusive coverage. Taking a break on the way to the flats is the HRM caravan that included two Ford panel trucks and the Evans, Starr & Alger Class B Lakester.

November 1950 Why are these men smiling? It could be because the trip to the flats was made easier with the addition of the company Ford panel truck. The world record board was yet to reflect the invasion of the hot rodders but Tom Medley and Wally Parks don't seem to mind.

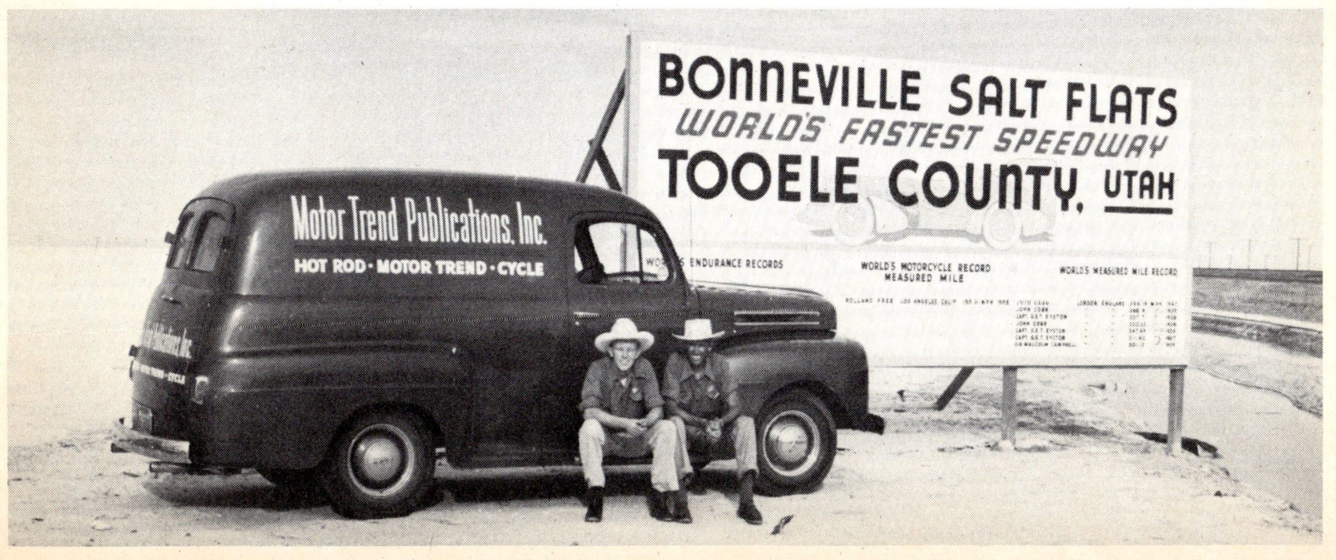

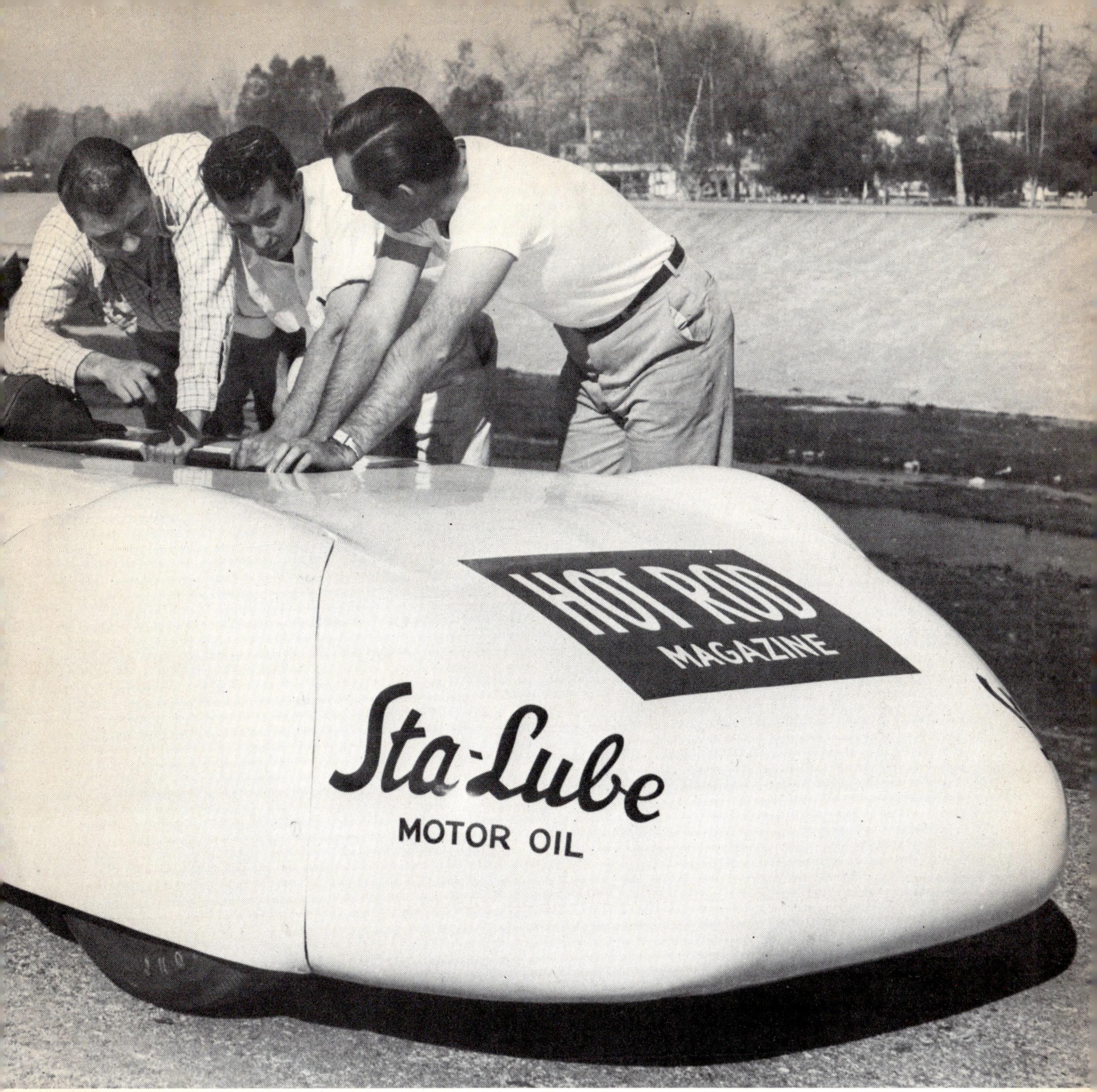
HOT ROD
MAGAZINE
Sta-Lube
MOTOR OIL

BONNEVILLE SALT FLATS
WORLD'S FASTEST SPEEDWAY
TOOELE COUNTY, UTAH
Motor Trend Publications, Inc.
HOT ROD · MOTOR TREND · CYCLE

May 1950 Built for the dry lakes, Ray Brown's '27 T was said to top the list for all-around originality in roadster construction. The stock Model T rails were boxed and kicked up at the back to house the rear-mounted '38 Ford V8 Sixty engine. The bright red "Sizzling Sixty" weighed in at just 1500 pounds.
(Photo: Tom Medley)

November 1951 The November Hot Rod of the Month was none other than Ak Miller's "projectile-nosed" '27 T roadster. The Mercury-powered rear engine "Missile" was designed by Art Ford and built in Ak's backyard garage. It featured a unique air ram for the carbs and aluminum wheel cover discs.
(Photo: Tom Medley)

October 1951 All over the nation, in backyard shops and garages, a "frenzied stir of activity" was taking place as hot rodders got ready for Bonneville. Among the shop scene activities reported on was a look at Howard Johansen's Crosley sedan-bodied car complete with top chop and "weird-shaped" nose.
(Photos: Tom Medley & Wally Parks)

April 1952 Another Alex Xydias So-Cal sponsored lakester was this class record-holding wing tank co-owned by Xydias and Dave DeLangton. The car's seasoned V8 Sixty engine ran previously in the famed "So-Cal Special." Controls for the car included hand-operated clutch and brake levers and hydraulic foot accelerator.
(Photo: Felix Zelenka)

July 1952 The Southern California Timing Association's 1952 season started with a 1.7-mile course at El Mirage dry lake. This roadster driver got an assist with his required safety helmet, from a saddle-shoe-wearing friend. (Photos: Joe Manriquez and Bill Burke)

August 1952 Ted Miller's cover '27 T roadster ran weekends at the dry lakes and turned 152 mph at the '51 Bonneville Nationals. Miller (right) and friend Bruce Brown worked on the '47 Merc powerplant equipped with a Weiand head and four Stromberg 97 carburetors. (Photo: Joe Manriquez)

September 1952 Harold Post's streamliner was built with an eye towards breaking the national speed record. The handmade aluminum body wraps around a tube chassis designed to accept either Mercury or Chrysler V8s. (Photo: Eric Rickman)

November 1952 George Hill's "City of Burbank" streamliner stops off in front of the HRM offices. The Hill/Bill Davis-built hot rod had just captured the International Class C speed record with an average of 229.77 mph. The engine used for setting the record was a 248-cubic-inch Mercury equipped with C and T Automotive heads and fuel injection.

December 1952. The "Impossible Car That Came True" was Chet Herbert's fiberglass-bodied streamliner. Herbert, working out of a wheelchair, survived a series of body construction delays, engine problems (one blew up on the dyno; another, a supercharged Miller was scrapped midway in favor of the eventual Chrysler) and hasty construction problems, to set the '52 National's second highest speed.

November 1952 Waiting for the starting signal at the '52 Nationals was the Waite-Bradshaw "C" Modified Roadster. The car turned 189.17 one way and a record two-way average of 186.09. (Photo: HRM Staff)

August 1953 A preview of 1953 Bonneville competitors found Mickey Thompson during construction stages of a new longer wheelbase coupe. When visited by HRM, Mickey was discussing the possibilities of running one Merc engine in conjunction with one Chrysler V8 in his dual engine record setter. (Photo: Eric Rickman)

February 1954 Billed as "The Most Fantastic Coupe!" on the February cover, Art and Lloyd Chrisman's rear-engined '30 Ford Model A coupe was built for the flats rather than the quarter mile where they, along with their father, had built a winning reputation. Chopped to the limit, the body featured a '40 Ford hood nose and full belly pan. Three engines were used at Bonneville, including a 258-cubic-inch Ardun-Merc that powered the car to a Class B record average of 160 mph. (Photo: Eric Rickman)

January 1953 Jointly owned and built by Dick Hubbard and Al Palamides and driven by Hubbard, this radical little Crosley sedan was equipped to accept three different flathead engines for different class competitions. (Photo: HRM Staff)

September 1954 Bonneville newcomer Craig Bowman's '34 Ford three-window coupe received a lot of attention during its construction due to its twin S.C.O.T. blown Merc setup. The front engine was mounted in the usual manner while the rear engine was mounted on radius rods and coupled directly to the quick-change rearend.

September 1954 HRM's annual Bonneville preview showed the shape of things to come and the shape of Gene Thurman's radical coupe was needle-nosed. The car featured a full roll cage and stood under four feet tall.

November 1955 The A.A.A. Contest Board took over the Bonneville flats to conduct their National and International speed trials. Jim Simpson came to the A.A.A. with a hand-made Italian O.S.C.A. designed by E. J. Fronteras. Forty new National and International records were set by drivers Tony Bettenhausen and Marshall Lewis in endurance categories on the 10-mile circular race course set out on the flats.
(Photo: Bob D'Olivo)

May 1954 "A winner first, now a classic" is how HRM Managing Editor Bob Green described Alex Xydias' Double Threat Coupe. The chopped Ford coupe saw double duty on the lakes and also "cleaned house" at the Pomona drags. The 258-cubic-inch Mercury V8 ran with a GMC 471 blower for the flats but shed the puffer for the drags.

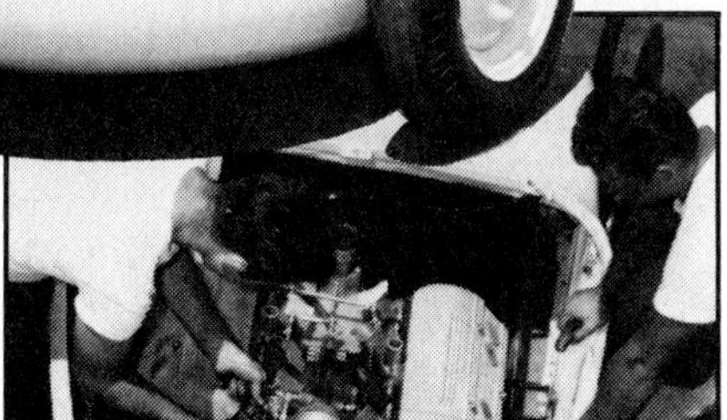

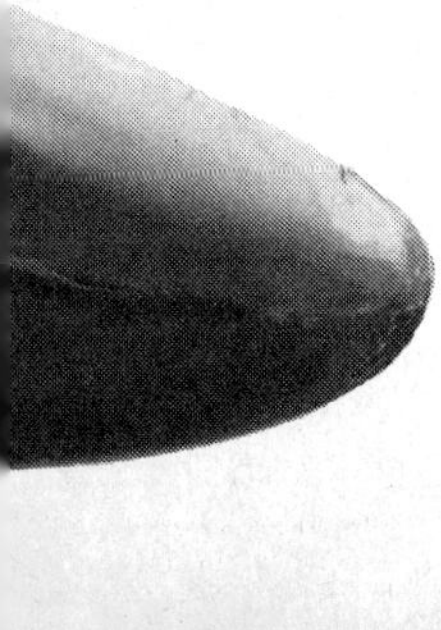

September 1955 Among the cars "Gunning For 300!" in Ray Brock's technical preview of Bonneville was Tom Ruddy and Marty Weinstein's sleek '27 T roadster. The Ardun-Merc-powered rear-engined screamer featured an enclosed front axle/wishbone design created by Chuck Porter's Body Shop.

September 1957 The Southern California Timing Association celebrated it's twentieth anniversary in 1957. Approximately 95 cars showed up for their El Mirage speed trial where the timing crew, headed by J. Otto Crocker (left), held reign. (Photo: Wally Parks)

October 1956 Carl Borgh's GMC-powered Bonneville roadster was said to be "worth its salt." The 1925 Chevy-bodied car was built originally by Spurgeon and Givonne and held the Class A El Mirage record in '48-49. Carl moved the 292-cubic-inch GMC behind the driver in 1955. (Photo: Bob D'Olivo)

March 1958 Fred Esser and Howard Leever's Model A-bodied roadster was an excellent example of quality workmanship. Doug Webster was responsible for the tube chassis and setup of the front and rear torsion bar suspension. Its '49 Olds powerplant was equipped with Hunt-Vertex mag, Hilborn injection, Herbert cam and ported and polished 10:1 compression '54 heads. (Photo: Eric Rickman)

February 1958 "The Thinnest Streamliner" measured only 25 inches across tire-to-tire. Owned by John Vesco and driven by Jim Dinkins, the streamliner carried a 116-inch wheelbase. Power came from a de-stroked 182-cubic-inch Ford four-banger with a Riley overhead conversion and John's own fuel injection. The tube chassis was fully suspended with leaf spring and tube shock front and flat leaf torsional radius rod/motorcycle tube shock rear. (Photography: Bob Hardee)

September 1959 Mickey Thompson was seldom out of the pages of HRM for long. He earned his reputation as one of the world's best known hot rodders with mind-boggling creations such as his four Pontiac V8-powered land speed record challenging streamliner. The front two engines faced backward to drive the front wheels, while the rear two were linked to the rear wheels. The sleek aluminum body on the rigidly suspended, unsprung chassis measured 19 feet long.
(Photos: Eric Rickman and Lester Nehamkin)

November 1958 The tenth running of the Bonneville Nationals saw speeds of 200 mph becoming commonplace. A Mercedes 300SL roadster was entered by Bill Scace. Sans windshield and with the addition of wheel discs, headrest and canopy, the car set a new sports car class record of 143.191 mph.

December 1959 While the streamliners and roadsters garnered most of the attention, HRM still reported on the efforts of the less exotic Bonneville classes. Winning the "A" Gas Coupe-Sedan class in 1959 was this Williams Automotive-sponsored Mercury, powered by a twin-McCulloch blown 390-cubic-inch '56 Lincoln.
(Photos: Bob D'Olivo and Ray Brock)

April 1960 The jet rodding age was ushered in by Ak Miller, Nathan Ostich, Ray Brock, and Allan Bradshaw, shown gathered around the General Electric J47-19 Turbo Jet engine of "Doc" Ostich's land speed record machine. Ostich hoped the car would be the first land vehicle to break the 500 mph mark. (Photo: Art Streib)

January 1964 Bonneville attracted the national media exposure but HRM also continued to report on the SCTA action at El Mirage. Trying to break into the exclusive 160 Club in '64 was A/Street Roadster recorder holder Tom McMullen. Helping to tune the supercharged 327-powered '32 highboy was Rose Gennuso. Tom barely missed surpassing his 157 mph record this time out. (Photo: Tex Smith)

August 1963 Walt Arfons was still trying to break the 400 mph Land Speed Record barrier with his Goodyear-sponsored Wingfoot Express. The 25-foot-long open-cockpit streamliner was powered by a J46 Westinghouse turbojet engine.

December 1963 Skimming along the ground just 2½ inches above the salt was a favorite pastime of Charles Markley, driver and owner with his brother Bob of the "Impossible Tank." The de-stroked '55 Plymouth-powered lakester was built from a modified P-38 aircraft belly tank and turned over 280 mph, once thought unattainable for open-wheel cars.

October 1963 At age 26, Craig Breedlove became the fastest man on wheels and the first to break into the 400-mph club. His "Spirit of America" jet-powered three-wheeler stood 10½ feet tall at the tail fin and measured 39½ feet in length. It used 57 gallons of Shell turbine-type jet fuel for each one way, full throttle pass down Bonneville's long black line.

January 1965 After taking photos of Craig Breedlove's unfortunate Spirit of America ditch plunge, Eric Rickman paused for a photo of himself. Breedlove had established a run of nearly 526 mph before losing control of his record-holding three-wheeled Land Speed car.

November 1963 Bonneville '63 saw 169 entries vying for 72 established class records. One of the littlest cars on the flats was Allan Richards' 2.8 cubic inch "whatsit" that broke the spectators up when its two-stroke Garelli engine "screamed commands" at the car's bicycle tires. Its top speed crowded 55 mph. (Photo: HRM Staff)

June 1965 HRM Ad Manager Bill Burke was saluted in the magazine in 1965 as "Mr. Prolific" for his contributions to hot rodding and the flats. Bill was credited with creating the first wing tank in 1946. The original lakester sported a Thickston-equipped Ford flathead. This car was followed by a series of record setters that continues into the pages of HRM.

July 1965 Art Arfons, long-time hot rodder and firm believer in big speed with big engines, shocked the world by running 536 mph in his home-built jet-powered "Green Monster." The question posed in July '65 was, could he break 600 mph and eventually the sound barrier. (Photo: Lester Nehamkin)

March 1964 Bob Herda used his experience as Chief Aerodynamics Engineer at Hiller Aircraft to help him design his record-breaking blown Chrysler-powered Herda-Cagle streamliner. Frontal area on the car was a low 11.4 square feet. (Photo: Eric Rickman)

September 1965 El Mirage SCTA action continued with a lot of radical entries, such as Nolan White's fiberglass A/Sports Racing car. High point man in the 1964 season, the Nolan creation was powered by a blown Chevy and reached speeds of 197 mph. (Photo: "Tex" Smith)

January 1969 Future HRM editor Lee Kelley set the C/Production class record with a two-way average of 166.133 mph at the rain-delayed 1968 Bonneville Speed Week. The Dave Mauer engineered Olds Cutlass was a multi-purpose HRM project car, which was put into action on the salt to compete against a three car Car Craft Magazine C/Production record assault. Kelley/Mauer beat all comers even though it was Kelley's first time ever on the salt. (Photo: Eric Rickman)

February 1969 Labeled as an "engineering exercise" Autolite Division of the Ford Motor Company showed up on the flats with the 20-battery-powered "Lead Wedge." The basic wedge concept was borrowed from the Granatelli turbines. The batteries drove a General Electric industrial motor rated at 40 horsepower at 8000 rpm. To reach the car's record speed of 138, the motor was rewired to give 120 horsepower on 90 kilowatts of juice.

April 1971 Mark Dees' "Repetitious Roadster" was so named because it just kept setting records. The fiberglass American Austin body, with a sectioned '34 Chevy nose, covered a Bob Joehnck-built 427-cubic-inch Chevy. The car ran 200-plus in '68, upped the record in '69 and broke the 220 mph mark at Bonneville in 1970. (Photo: Eric Rickman)

February 1971 The World Land Speed Record was pushed above the six century mark to 622.407 mph by the "Blue Flame," a four-wheeled liquefied natural gas-powered rocket driven by Gary Gabelich. Actually the second vehicle to break 600 (Breedlove ran 600.601 a year earlier), Gary said the Blue Flame's next goal would be to crack the sound barrier. (Photo: Eric Rickman)

November 1976 The Tony Nancy-driven Leggitt & Mead "Spirit of '76" was limited to a 284-mph run in '76 but still managed to set fast time of the meet and win HRM's top time trophy for the third time in a row. (Photo: Gray Baskerville)

January 1979 The streamliners were still going strong in '79. A good example was the Immerso T-bird. The brilliant orange and black beauty was turbine-torqued. It set a class record in '79 of 264.903 mph. (Photo: HRM Staff)

January 1980 Bonneville, the ultimate goal for the faithful, continued to draw loads of entrants and both new and familiar cars. Van Prothero upped his A/Production class record to 219 mph. The stock-bodied '67 Camaro was powered by a 500-cubic-inch Chevy big-block.
(Photo: Gray Baskerville)

March 1980 The world's fastest stock-bodied Corvette was the only way to describe the Bob Kehoe/Duane McKinney/Gale Banks' "Sundowner." Based on a '68 Vette body, the car got its record-setting speed from the Banks-built 460-cubic-inch Chevy.
(Photo: Gray Baskerville)

The Hot Rods

In the beginning, there was no distinction between hot rods and street rods. *All* old cars with modified drivetrains were considered hot rods. The term hot rod was really a rather derogative term, one closely associated with words like "juvenile delinquent" and "hell-raiser." HOT ROD Magazine and groups like the National Hot Rod Association worked long and hard to make hot rods respectable. Street rod, however, is the contemporary term.

Some cars like '40 Ford coupes and '36 Ford roadsters will be found in both the street rod chapter and the street machine chapter. Depending on the year and modifications, many cars can fit into more than one category. For instance, what is considered a very old car and excellent street rod material like a '40 Ford, was in 1948, just a moderately used street machine. It is interesting to note how much street rods have changed and, at the same time, have remained very similar to their 1948 counterparts. The venerable flathead engine is all but gone, and many bodies are now fiberglass reproductions, but the classic hot rod look still prevails. Trends like fenderless and full-fendered cars come and go, but the fun and excitement of a fine street rod has endured as long as HOT ROD Magazine.

May 1963 One of the best known hot rods of all time was Tom McMullen's classic '32 Ford highboy. The bright flames were done by Tom and pinstriped by Ed Roth. The car's blown small-block Chevy pushed it to times of 11.59 seconds at 127 mph in the quarter mile and 151 mph at the dry lakes. Tom came from La Habra, California, and wrote many freelance articles for HOT ROD before establishing his own publishing company. (Photo: Eric Rickman)

February 1948 This '32 Ford roadster belonged to Ed Stewart (right). Ed was a member of the San Diego Roadster Club and was responsible for those San Diego dropped axles.

February 1948 The cover car and Hot Rod of the Month was this '32 Ford roadster which belonged to Keith Landrigan. The engine was a '38 LaSalle V8. The car was timed at 115 mph at the dry lakes. The mufflers were encased in the ends of the large pipes that extended from the headers.
(Photo: Lee Blaisdell)

February 1948 Art McCormick, his wife Ellena, and son Gary are shown at the turnout for the Hot Rod Exposition display. This meeting was held at the new Lincoln-Mercury plant in Los Angeles. The body of the car is a '29 Model A.

April 1948 The engine in Walter Rose's '27 T roadster was rather unusual—a supercharged Cord V8 with a Winfield camshaft. The doors were welded shut and cut down.

June 1948 The May cover car belonged to Ed Iskenderian, the well-known camshaft grinder. The car was first built in 1940. The engine was a '32 Ford with Maxi Overheads which propelled the T-bodied car to a top speed of 120 mph at El Mirage dry lakes. The original purchase price of the '24 Model T was $4.00.

February 1951 A popular form of outing for street rod owners in the early Fifties was the reliabilty run, which was a time and distance navigational rally. This one was sponsored by the Pasadena (California) Roadster Club, so it started and ended at the famous Rose Bowl. The time for the course was a little over three hours and 100 cars entered the event. (Photo: Eric Rickman)

March 1950 The winner of the "America's Most Beautiful Roadster" trophy at the 1950 Oakland Roadster Show was Bill NieKamp's sleek, track-nosed Model A. The car was built from four different '29 Model A bodies on a frame from a '27 Essex. The engine was a '42 Mercury flathead with Evans cylinder heads. The car is still in existence after being restored by its present owner, Jim Jacobs. (Photos: Eric Rickman, Tom Medley)

May 1951 At the '51 Oakland Roadster Show The NieKamp car was runner-up for the A.M.B.R. title. (Photo: Eric Rickman)

May 1950 Don Ferrara bought his Model A roadster right after high school and spent seven years getting it into good enough shape to appear in HOT ROD. The engine was from a '37 Ford; equipment included a Offenhauser manifold, three Stromberg 97's, Edelbrock pistons and Cyclone cylinder heads. The car was painted berry red with a maroon and eggshell plastic interior. Although built as a street car, Don drove the car to a top speed of 121 mph at the dry lakes. Tom Medley shot the photos with a 4x5 Speed Graphic, the first company camera, originally owned by R. E. Petersen. (Photo: Tom Medley)

July 1950 Louis Banta and Sal Macchia were partners in a Los Angeles Mobil station and a sharp '27 T roadster. The car was painted bright vermilion and white to match the Mobil colors. The engine was a '42 Merc with Weiand heads and triple carburetor intake manifold. The small chrome pump just below the windshield post was a hand-operated fuel pressure pump. (Photo: Tom Medley)

April 1951 The first HOT ROD cover with a color photo featured Jack Morgan's channeled '34 Ford roadster from Santa Ana, California. The bright yellow car had a distinctive look with its lack of fenders and use of a '37 Ford truck grille. The exhaust headers on the '46 Merc engine fed into four-inch diameter exhaust pipes that used slip-in mufflers for road use. (Photo: Tom Medley)

October 1950 This '32 Ford roadster was built by Mac Schutt of Bel Air, California, for cruising around town and entering car shows. The body was channeled over the frame five inches in front and six inches in the rear. The engine was the very popular '42 Mercury flathead. The windshield was chopped and new safety glass was installed. The body was painted maroon and the engine white. (Photo: Tom Medley)

January 1951 The cover photo was of Kenny Smith's roadster which was featured on KTLA's (a local Los Angeles television station) *City At Night* program. The location was the 1950 Los Angeles Motorama. The live program also featured interviews with the owners of the So-Cal Special Bonneville streamliner, Alex Xydias and Dean Batchelor. (Photo: Tom Medley, Art Direction: Al Isaacs)

May 1951 This group shot from the 1951 Oakland Roadster Show covers only a small section of the big show. Fenderless roadsters were the predominant style of car at the show. (Photo: Eric Rickman)

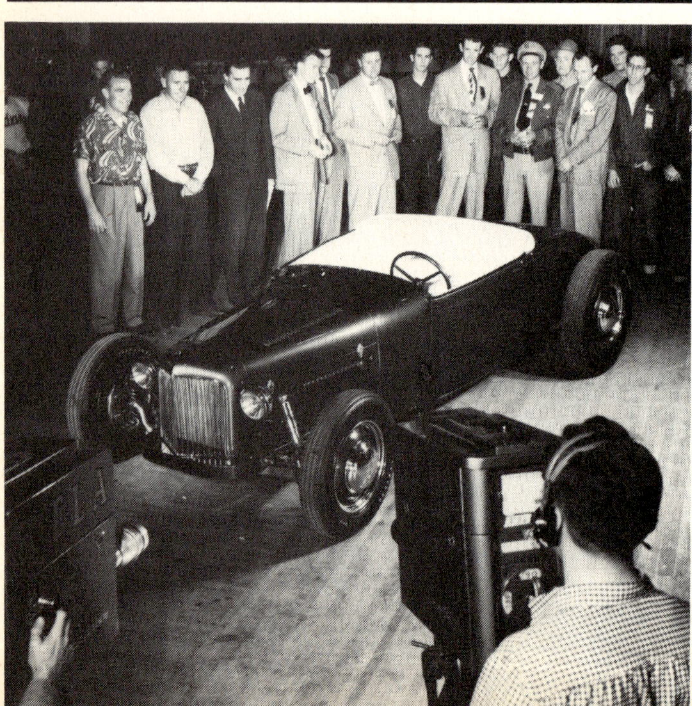

November 1951 Don Williams of Lynwood, California, built a '32 Ford coupe that was about as low as you could go. The bright red car was chopped five inches, channeled over the frame, the front axle was dropped and the rear of the frame was kicked up seven inches. The tall 16-inch tires and wheels furthered the low look. (Photo: Felix Zelenka)

October 1951 Vaughn Vance built his '29 Ford roadster pickup for street and strip action. The body was mounted on a '32 Ford frame and a '32 grille was used with a custom hood. The exhaust system went under the vehicle and re-emerged at the base of the pickup bed to exit at the tailgate. (Photo by Pete)

August 1951 Bob and Jerry Pederson of Los Angeles chose the seldom used '32 Ford Tudor sedan body as the basis of their highboy street rod. The '47 Mercury engine pushed the car to a top time of 96 mph at the Santa Ana dragstrip. The color combination was chrome yellow and lemon yellow with a red engine block. (Photo: Felix Zelenka)

January 1952 The Hot Rod of the Month was Roy Desbrow's chopped and channeled '32 Ford pickup. The truck was painted '50 Mercury cream by Tom Noel, who also did most of the bodywork. A '41 Ford bumper was used in front and nerf bars were used in back. (Photo: Felix Zelenka)

May 1952 Bud Parham of Portland, Oregon, built his street rod out of a 1925 Model T. The frame was a combination of a Model A and a '31 Chevy. A grille from a '32 Ford was used. The front fenders were made from spare tire covers. Bud was president of the Columbia Timing Association. (Photo: Robert Hegge)

June 1951 Rather than just photographing Dave Mitchell's highboy roadster pickup, HOT ROD Editor Wally Parks and Motor Trend Editor Walt Woron road tested the car. Zero to sixty was reached in 9.7 seconds thanks to the Olds V8 powerplant. The car stopped in 98 feet from 45 mph. (Photo: Felix Zelenka)

May 1951 Among the fine cars entered at the 1951 Oakland Roadster Show was Dick Kraft's track-style T. The car was around for many years and was later known as the Highland Plating Special. (Photo: Eric Rickman)

May 1952 The cover car and Hot Rod of the Month belonged to Dick Flint who was a college student attending Los Angeles State College at the time. The special bodywork including a contoured belly pan was performed by Valley Custom Shop. The dash included an instrument panel from an Auburn and a SCTA plaque certifying that the car turned 143.45 mph. (Photo: Felix Zelenka)

March 1953 Gordon Potter of Hawthorne, California, had his Model A roadster and later, an early Thunderbird featured in the pages of HOT ROD. Potter's cars were heavy on horsepower with Chrysler Hemis being used in both cases. The roadster reached a top speed of 153.51 on straight alcohol at Bonneville, and 145 mph was reached on gasoline. When not competing, the car was used on the street. (Photo: Felix Zelenka)

June 1953 Frank Rose's '27 T had many distinctive features including a frame made from chrom-moly steel tubing, a full belly pan and hand-formed fenders. The rear fenders and deck lid could be quickly removed because they were held in place with Dzus fasteners. The body work and black lacquer paint were handled by Jack Hageman of Castro Valley, California. (Photo: Eric Rickman)

November 1953 The most noticeable features on Frank Mack's Farmington, Michigan-based '27 T were the 1927 Jordan headlights. Brass knock-off spinners held '47 Hudson hubcaps to the 15-inch Mercury wheels. Frank did all the work on the car including the black lacquer paint. (Photos: Medley and Hitchingham)

October 1954 Ak Miller has appeared in technical and feature articles in HOT ROD for over three decades. Always the innovator, his Olds-powered '32 Ford roadster used A-frame suspension from a '40 Chevy.

October 1954 Al Knoll's roadster was billed as a real "Chute Cutie" because of its dual functions as a street rod and drag machine. The Bell, California, car was comprised of a '28 Ford body, '32 frame, '32 fenders and running boards and '48 wheels. Al made the headers for the '38 Ford engine from steel bed posts. Taillights were from a '41 Chevy, and the headlights were Appleton units. (Photos: Tom Medley and Dick Day)

October 1955 The car that revolutionized the look of street rods was on the October cover. Norm Grabowski's T roadster pickup set the style for what became known as the "Fad T." The engine was from a '52 Cadillac and the transmission came from a '39 Ford. The car later gained national exposure as a feature car in the TV series, *77 Sunset Strip.* (Photo: Bob D'Olivo)

August 1955 This trio of street rods all came from Indianapolis, Indiana. The lead car belonged to Jerry McKenzie and started life as a '27 T roadster pickup. Jerry built his turtledeck from the top of a cab of a '39 Dodge panel truck. The car in the middle was also a '27 T which belonged to Bill Robertson. Both roadsters used flathead power. The owner of the '34 coupe wasn't identified. (Photo: Eric Rickman)

September 1957 When traditionalists think of '40 Fords and flame paint jobs, Bob McCoy's hot '40 sedan usually comes to mind. Ray Cook of San Diego, California, was responsible for the wild red, yellow and gold flames. A dropped front axle gave the car that just right rake. Ray Cook also made the nerf bars. (Photo: Bob Hardee)

December 1955 The December cover car belonged to Don Van Hoff of Los Angeles. The top was chopped 2½-inches before the car was painted Royal Maroon. The roof insert was done in white to match the wide whitewall tires. (Photo: Eric Rickman)

August 1957 Besides Norm Grabowski's T bucket, probably the next most influential T was the one belonging to Tommy Ivo. Tommy's Buick-powered Titan red bomb was as fast as it was good-looking. The car was capable of 12-second 117-mph quarter-mile times while picking up Best Appearing awards. The car later received more exposure under the ownership of Bill Rowland. (Photo: Bob D'Olivo)

October 1957 Lloyd Bakan's '32 Ford coupe from Los Angeles always seemed to draw plenty of attention from pretty young girls. A swimming pool shot with two bathing beauties was used on the cover. The top was chopped five inches. The front fenders were cycle fenders and the rear fenders were the stock items used without running boards. Extensive use of chrome was made including the '54 DeSoto engine. (Photo: Eric Rickman)

October 1958 One of the most famous street rods of all time was the Ala Kart which was built by George Barris and owned by Dick Peters of Fresno, California. The highly modified '29 Ford roadster pickup twice won the Oakland Roadster Show title, "America's Most Beautiful Roadster." The radical rod used the rear section from a '27 T to give the A body a different look. The bed was handmade. The truck had air-lift suspension and extensive use of chrome. The paint consisted of thirty coats of imported Swedish Pearl. AMT made a very popular model of the Ala Kart. (Photos: George Barris)

December 1958 Good street rods have a way of coming back again and again. Skip Torgerson's '27 T was featured several times before (as early as 1950) when it was owned by Lou Banta and Sal Machia. Torgerson of Hollywood, California, spent two years and $1800 restoring the car. Improvements included a custom top by Tony Nancy. (Photo: Eric Rickman)

March 1958 Bob Smith of San Diego and Bonnie Smith posed with Bob's '27 T on the March cover. The T was sectioned six inches and channeled over the '32 Ford frame. The engine was a 256-cubic-inch Mercury with an Evans manifold and finned cylinder heads. (Photo: Bob Hardee)

April 1962 Talk about a street rod with lasting power—Dick Scritchfield's '32 highboy first graced the pages of HOT ROD in October, 1948 when it was owned by Bob McGee. Dick bought the car in the Fifties and still owns it. The car has gone through a variety of changes over the years, but it has always maintained that traditional hot rod look. (Photo: Eric Rickman)

January 1963 Clark Christie of El Segundo, California, built a beautiful example of a classic street rod. The car was simple, devoid of doodads and expertly detailed. Chromed, reversed wheels with Merc hubcaps were used on the '29 roadster. The engine was a small-block Chevy backed by a '48 Ford gearbox. (Photo: Eric Rickman)

March 1959 The scallop trend even found its way onto street rods like John Rasmussen's March cover car. The Westchester, California, A was painted Bahama Blue with white scallops. The interior color scheme matched the exterior. Individual bucket seats were used. The car also had a one-piece removable top with T-bird portholes. (Photo: Eric Rickman)

December 1959 The '58 Corvette engine with Hilborn injection made Bill Strickland's Niantic, Connecticut-based '24 T a real stormer. The metal-bodied car turned 114 mph the first time at the drags. (Photo: Ed Eaton)

October 1960 Joe Cruces had his '21 Model T coupe and his attractive wife on the October cover. The Vacaville, California, rod was detailed to the limit. Joe spent a lot of time underneath the car polishing all the chrome suspension and exhaust components. Red and white pleated naugahyde covered the roof and running boards. A '48 Mercury engine was used, and the tall T rolled around on Buick Skylark wire wheels and Bruce whitewall slicks. (Photo: Dave Cunningham)

August 1960 Norm Grabowski followed his famous Kookie T with a more conservative 1925 Model T tub. Both Norm and the car appeared with Mamie Van Doren in the epic movie, *Sex Kittens Go to College.* The plush diamond tufted seats were done by Bill Colgan, while Valley Custom did the body work. (Photo: Eric Rickman)

January 1960 One of the better-known Eastern cars belonged to Bill Neumann of White Plains, New York. Shortly after this photo feature ran in HOT ROD, Bill and his '31 roadster moved to California where he became editor of Rod & Custom magazine. The bright red body was channeled 9½ inches. The engine was a '58 Chevy with an Offenhauser manifold. (Photo: Gareth Neumann)

August 1960 Long-time hot rodder Tony La Masa of Los Angeles built this nimble '32 roadster to show what he thought was an ideal American sports car. The vivid green roadster was used by Ricky Nelson on the *Ozzie and Harriet* show. Underneath the louvered hood was a stock '56 Corvette engine mated to a Lincoln Zephyr transmission. (Photo: Eric Rickman)

June 1961 Rod & Custom Magazine staffer Neal East entered his '32 Ford roadster (the former Bill Woodard car) in the '61 Winternationals car show. The display included lots of promotional material for R&C. The undercarriage and '56 Chevy V8 were highly chromed but that didn't stop Neal from driving the car daily. (Photo: PPC Photographic)

December 1961 Dave Shorter of Vancouver British Columbia, turned a $15 Model A into a show champion. He mounted the body on a '32 Ford frame and fenders. The engine was a blown '55 Dodge. The transmission was from a Packard and the rear axle was from a Mercury. (Photo: Peter Sukalac)

July 1961 Clarence "Chili" Catallo from Dearborn, Michigan, revamped his '32 Ford coupe four times, ending up with a wild show rod. The Alexander brothers shop in Detroit performed the special body work including the custom grille. The powerplant was a blown '56 Oldsmobile. George Barris chopped the top and applied the translucent Oriental Blue and Pearl paint. The whitc naugahyde interior was stitched up by Larsen Upholstery. The front tires featured whitewalls on both sides. (Photo: Eric Rickman)

October 1961 The big attraction on Howard Mitchell's '30 Ford coupe was the wild engine. Underneath a maze of chrome plumbing was a 454-cubic-inch '57 Lincoln. Also buried in the engine compartment was a McCulloch blower. The LaGrange Park, Illinois, car had won over 160 show awards at the time of the story. Chrome plating was everywhere on the car including the entire frame. Howard claimed an investment of fifteen thousand 1960 dollars in the car. (Photo: John deCampi)

March 1962 Dick Kraft first built this '25 T in the late 1940s. In the 1960s, Richard and Gary Seiden rebuilt and modernized the car, calling it the Highland Plating Special. Gary was seen showing Diane Terry how to tighten the bolts on the blower manifold of the '46 Merc engine. The bright red track-style roadster made the magazines again in the 1970s when Ron Weeks revamped it. (Photo: Eric Rickman)

April 1962 The Jackman brothers, Tom and Harry, picked a rare '32 Ford Sport Coupe for the basis of a wild show rod. The car was painted Candy Wild Cherry with a silver naugahyde interior. The floorboards were clear plexiglass so that the extensively chromed chassis could be seen. The underside of the El Cajon, California, car was a mass of chromed exhaust pipes. The engine was a '57 Ford 312-cubic-inch model with Holley dual quads. The wheels were polished Halibrands. (Photos: Bob Hardee)

May 1962 The 14th annual Oakland Roadster Show was filled with fine street rods as this partial view of the show floor demonstrates.
(Photo: Pete Biro)

May 1962 The proud man standing next to the towering trophy awarded at the Oakland Roadster Show for the title "America's Most Beautiful Roadster" is famed customizer George Barris. George won the nine-foot-tall trophy with his sparkling, Metalflaked '27 T.
(Photo: Pete Biro)

December 1962 Jerry Volavka's '31 Ford sedan from Littleton, Colorado, was run under the title "Carriage for a Queen" and pretty Carrlyn Ward was the royalty in the story. The car was painted Mother-of-Pearl White with Candy Purple faded along the fender edges. (Carrlyn's bathing suit was a matching purple.) The interior was done in diamond-tufted purple velvet, and the horn button was a '59 Cadillac taillight lens. The engine was a well-chromed '56 Buick.
(Photo: Tony Spicola)

RIVIERA
BY BUICK
FORD CUSTOM CAR CARAVAN
NO UNECESSARY CONVERSATION WITH GUARD
TEEN AGE FAIR

July 1963 Walt Kaline of Whittier, California, owned a chrome plating shop, and it showed in his dazzling '24 T bucket. The engine was a rarely used '56 Pontiac unit with a Hydramatic transmission. The cars in the background and the pearl green roadster were from the Ford Custom Car Caravan.

(Photo: Eric Rickman)

October 1963 When the L. A. Roadster Club and the Bay Area Roadster Club from San Francisco got together for their annual North-South Run, the scenery was spectacular. At this meeting in Fresno the turnout of fine roadsters included the well-known Dick Flint track-nosed car which appeared on the May '52 cover of HOT ROD. (Photo: Eric Rickman)

WARNING
NO LIFE-GUARD ON DUTY
HOT ROD
Magazine
CALIFORNIA
JFS 004
CALIFORNIA
BYZ 193

September 1963 HOT ROD tried to cover the back-to-school clothing scene in an article entitled, "Fancy Campus Cloth For Hip Guys." Mickey Thompson wheels were used on John Monteiro's '27 Model T touring, and power came from a Chevy small-block of 1957 vintage. The car was also featured in the October '63 issue. (Photo: Eric Rickman)

May 1964 Candy Burgundy paint and brass-plated wheels and a pearl white Naugahyde interior created an interesting contrast on Gene Chan's '23 T (top). The engine was an early Chrysler Hemi with an Edelbrock eight-carb manifold. The rear turtle deck was covered with louvers on the Visalia, California, cruiser. (Photo: Ron Taylor)

June 1964 Chauvin Emmons' '23 T from Phoenix, Arizona (top), had a distinct Indy flavor thanks to its special nosepiece. The engine was a '58 Olds with Hilborn injectors. Halibrand wheels were used all around. Corky Russnak was the pretty passenger. (Photo: Bill Turney)

February 1964 Cliff White of Santa Ana, California, chose a rare '31 Model A roadster pickup for his trips with the L. A. Roadster Club. Burl Wilson applied the Naples Orange paint. Power came from a '60 Corvette engine with a Duntov camshaft. Donna Cantrell came along for the photo session at Newport Dunes in Newport, California. (Photo: Eric Rickman)

July 1964 The beautiful Garden of the Gods in Colorado Springs, Colorado, provided the setting for Verlin Lawson's '32 Ford coupe. Denise Williams also added to the scenery. An Olds engine and a '37 LaSalle transmission made up the power train.

(Photo: Tony Spicola)

July 1965 Sedan deliveries were just starting to gain popularity as street rods when the feature was run on Michael Ray's '28 A from Sepulveda, California. The bright red hauler was built for less than $1000. Power was from a venerable flathead Ford of 1948 vintage.

(Photo: LeRoi Smith)

July 1964 Ted Wingate's two '32 Ford roadsters graced the July cover of HOT ROD and were featured in the August issue. The red drag car was clocked at 109 mph at 12.62 seconds in B/SR class with a weight of 2980 pounds. The engine was from a '62 Corvette. The channeled street car was powered by a '56 Buick engine mated to a '38 Buick transmission. Ted resided in South Pasadena, California, and was a member of the L. A. Roadsters Club.
(Photo: Eric Rickman)

January 1966 Dick Patterson had more than ample power for his chopped and channeled '31 Model A coupe. The 312 Thunderbird engine employed two McCulloch superchargers for additional horsepower. The wild interior of the Rialto, California, car was made out of padded and upholstered pieces of exhaust tubing.
(Photo: Bob Wagner)

TEMPORARILY
OUT of SERVICE

December 1966 Dick Rundell's 1928 Ford Phaeton made a great garage scene cover with its Hilborn-injected 301-inch Chevy small-block. The Long Beach, California-based car was painted Candy Orange acrylic by Cerney's Auto and contrasting black Naugahyde was stitched by Ed Martinez. (Photo: Eric Rickman)

February 1966 Dick Bouthillier was the owner of this '29 Model A sedan (below), although Russ Meeks originally built the car. Both men were residents of Portland, Oregon. The car was extensively detailed with polished brass on the wheels, undersides of the running boards and fenders, and the firewall. A '56 Corvette engine was topped by an Offenhauser manifold with three Rochester carburetors. (Photo: Peter Sukalac)

May 1968 The tall T look was popular in the Sixties and Jay Kovarsky's '26 T coupe (right) was a perfect rendition of the style. The polished 4-71 blower and the American mags with no front brakes added to the look. A dropped '31 Ford front axle and a '56 Chevy rearend made up the suspension. Home for the street T was Venice, California. (Photo: Don Emmons)

August 1968 If one photo could depict the California surf and car scene it was this shot of Meril Smith's 1928 Ford Woody. Sun, sand, surf, a bright red hot rod and a couple of cute girls (we assume that a Beach Boys song was playing on the radio) were a dream setting for more than a few cute guys. The Woody was powered by a 283 Chevy and a Powerglide automatic. The project was completed in eight months at a cost of $800. (Photo: Eric Rickman)

August 1963 Easily the best known HOT ROD project car was the XR-6 which was owned and conceived by HRM Associate Editor, LeRoi "Tex" Smith. Steve Swaja was responsible for the car's futuristic design and Gene Winfield, George Barris and Tony Nancy were among the many talented craftsmen that helped construct the car. The car's name came from the fact that it was an experimental roadster powered by a slant 6-cylinder engine.

Among the awards won by the famous roadster was the title "America's Most Beautiful Roadster" at the Oakland Roadster Show. The car was available in model form from AMT. (Photo: Eric Rickman)

July 1965 Elements of photographer Andy Southard, Jr.'s work always seem to include excellent photos, outstanding cars and beautiful women. Andy's personal '29 A roadster with Patricia Pere perched on the running board was no exception. The car was painted Rustic Maroon and pinstriped by Andy. The engine was a '62 Chevy 327 that put out 360 hp. Andy and the car spent their time in Salinas, California. (Photo: Andy Southard, Jr.)

August 1962 The first car to be featured on an inside color page in HOT ROD was Warren Swartzendruber's Model T pickup. The Roman Red rod was powered by a 283 Chevy with four Stromberg 48s on a Weiand manifold. Warren was from Upland, California. (Photo: Eric Rickman)

July 1976 When HOT ROD's street rod authority Gray Baskerville made his list of the World's 10 Greatest Rods he naturally included Andy Brizio's '23 T pickup. Known as Andy's Instant T, the car was a kit car that was turned into an Oakland Roadster Show winner. The psychedelic paint job was the work of Art Himsl. (Photo: Andy Southard, Jr.)

November 1978 Steve Group's idea of an outrageous street rod was to build a Fuel Altered for the street. Steve built the '23 T-bodied car himself including the blown 392-cubic-inch Chrysler Hemi that was set up to run on alcohol. The chassis was made from 1¾-inch tubing. The paint was Candy Blue with stripes in shades of red and orange. It's hard to believe, but Steve drove the car on a regular basis, stopping traffic and snapping necks everywhere he went. (Photo: Gray Baskerville)

July 1978 The cover photo of Phil Cool's '32 highboy cruising down the open road with his family by his side pretty well sums up what street rodding is all about. The Cloverdale, California, car epitomized the clean and mean look of hot rodding—more than enough horsepower from a blown 427 big-block and no extra frills to slow down the fun. The car was the recipient of the 1978 America's Most Beautiful Roadster title.
(Photo: Bob McClurg)

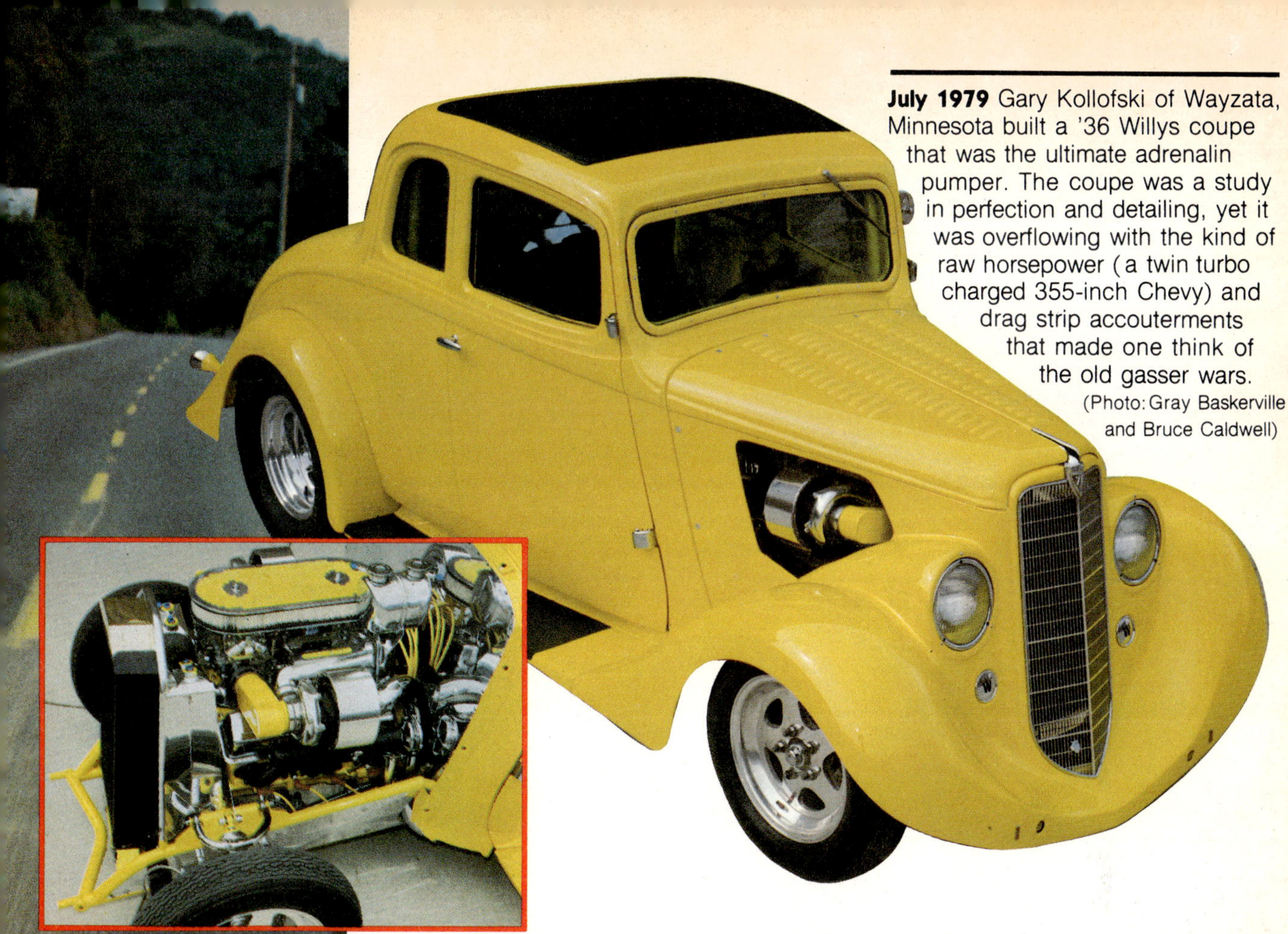

July 1979 Gary Kollofski of Wayzata, Minnesota built a '36 Willys coupe that was the ultimate adrenalin pumper. The coupe was a study in perfection and detailing, yet it was overflowing with the kind of raw horsepower (a twin turbo charged 355-inch Chevy) and drag strip accouterments that made one think of the old gasser wars.

(Photo: Gray Baskerville and Bruce Caldwell)

February 1981 The caliber of workmanship on street rods like John Corno's '29 highboy from Portland, Oregon is outstanding. The car is as close to perfect as craftsmen like John Buttera, Mike McKennett, Steve Davis and designer Harry Bradley could get. The street rods of the Eighties still look like the early cars, but the engineering and craftsmanship exceed that found on exotic luxury cars.

(Photo: Pat Brollier)

April 1978 John Siroonian of Fresno, California, founded the Western Wheel Company and used some of the profits from his labors to build a collection of the finest '32 Ford street rods in the world. One of his many beautiful open cars, a '32 Phaeton, was the subject of the April Rod Test. The engine was a small-block Chevy topped by a full set of Webers. Don Theilan was responsible for the flawless paint and bodywork.

(Photo: Gray Baskerville)

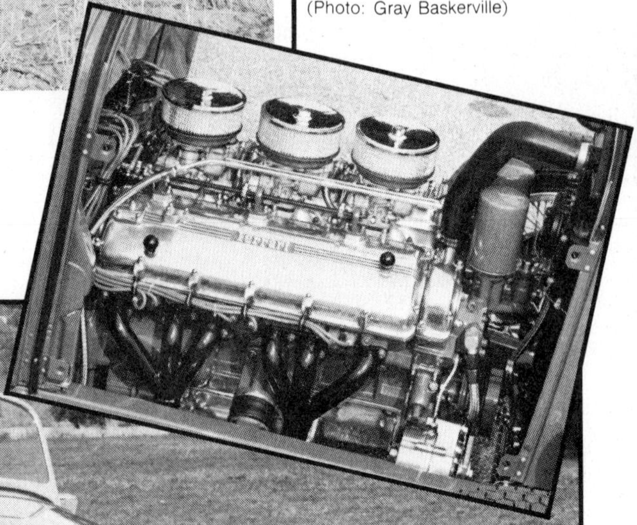

June 1979 If the two most popular engines for street rods were the small-block Chevy and the flathead Ford, probably the engine you would be least likely to see would be a Ferrari. Yet, Brian Burnett managed to put a Ferrari V12 into his '32 Ford roadster so beautifully that it looked like it came that way. Dick "Magoo" Megugorac was responsible for the construction of the four-inch stretched fiberglass Deuce. Brian is the son of HOT ROD's famed cutaway illustrator, Rex Burnett. The car won the 1979 "America's Most Beautiful Roadster" title. (Photos: Andy Southard, Jr.)

July 1981 Designer Thom Taylor got together with car builder extrodinaire, Boyd Coddington, to build the ultimate hot rod coupe for Vern Luce of Newport Beach, California. Coddington lavished the kind of attention and workmanship on the '33 coupe that had previously been reserved for open cars. The car looks so slick because of subtle changes to the body, like lifting the rear one inch, lengthening the wheelbase three inches and chopping the top three inches in front and 2¾-inches in the rear. The look of Vern Luce's coupe is classically simple yet the workmanship and execution are as complex as on any street rod ever built. (Photo: Bob D'Olivo)

The Lead Sleds

Nowhere in the sport of hot rodding was the diversification of interests more obvious than in the emergence of custom vehicles. The builders and owners of the "lead sleds" (so called because most early bodywork was done with lead, rather than with the newer plastic fillers) were seldom interested in racing or engine modifications, except for chroming everything that moved. Yet the customs had their place in HOT ROD and through its pages the word of the masters of the custom art became well-known. Barris, Winfield, Starbird and the Alexander Brothers were names that were just as familiar as Edelbrock, Weiand, Navarro and Iskenderian to the readers of HRM.

On one side of the lead sled picture were the milder street machines and rods, and on the

March 1953 This Barris-built 1951 Mercury was owned by Bob Hirohata. It featured chopped top, frenched '52 Lincoln taillights, full fender skirts with unique scoops and a single-bar floating-type grille. (Photo: W. G. Brown)

other side were the wilder experimental and show cars. For the ultimate in hamburger stand cruising, however, you couldn't beat a nose job and Carson top or a wild set of flames or scallops. Today full customs are as rare as a pair of full fender skirts or an Appleton spotlight, but their heritage can be seen in the styling and paint of the modern cruisin' machines. Styles may change, but for many hot rodders, Kustoms will always be King.

February 1952 Earl Bruce's smooth, chopped '40 Ford coupe was the product of 10 years of constant reconstruction. In its February '52 cover form, the custom featured filled rear quarter windows, rounded door top corners, remote control door latches and Swift red paint. The car was powered by an Eddie Meyer-built Mercury flathead. (Photo: Eric Rickman)

March 1953 As early as 1953, a young George Barris was showing HRM readers how to do their own customizing. Here, he showed how to fill in holes in the body that were left when chrome strips and ornaments were removed. Welding the holes was unacceptable because it caused warpage. The Barris method eliminated that problem. This method was a basic customizing tip that provided "valuable metal-working experience." (Photo: W. G. Brown)

January 1953 Roy Dunn's 1950 Ford Tudor was a major attraction at the 1952 Motorama. The body was shortened five inches, the wheelwells were cut out to conform with wheel radius and a wraparound single bar was used in the grille treatment. The parking lights were removed from the fenders and the headlights were frenched. The total cost of customizing, including new upholstery, was $1800. (Photos: W. G. Brown & Eric Rickman)

November 1953 Stanley Moore owned this '47 Lincoln Continental convert. He did all of the body work, which included partial dechroming, filling and molding the hood and installing '50 Olds "98" taillights. The engine compartment contained a modified Chrysler V8 and a log-type manifold that used four '53 Ford-six carburetors. (Photo: Thomas J. Medley)

May 1953 In the early Fifties, building or modifying a car's grille allowed the most freedom of expression of any of the tricks employed by custom car enthusiasts. HRM showed not only how to do it, but also provided examples of the latest trends.
(Photos: E. Rickman & J. Bailon)

September 1953 Jack Stewart's "Polynesian" Olds cover car was a restyled 1950 "88" Holiday. A four-inch-wide section was removed from the sides of the car, rear fenders were reshaped and boxed, air scoops for the rear fenders were fabricated, and the body was completely dechromed. Head and taillight rims were cut, perforated, welded and chrome-plated; the headlights were frenched, and a unique grille was fabricated. The final touch was "Orchid Flame" paint.
(Photo: Eric Rickman)

August 1955 Oliver Hines' 1933 Ford gathered most of its "speed-at-rest appearance" from its channeled body and sweeping imported grille. This $5500 glamor model was powered by a 250hp OHV Cadillac V8. (Photo: Joseph Farkas)

March 1957 Don Telen's "Meticulous Merc" began as a '40 stocker. A dropped front axle, raised rear crossmember and reversed spring eyes and 15-inch wheels lowered the coupe to a mere three inches above the ground. It featured a nosed hood, chopped top and molded fenders. To finish it off, the car was painted in Frosty Grape lacquer. Don's only complaint, "Tire changes are a major problem." (Photo: Pete Sukalac)

April 1956 Leonard Thompson's five-in-one custom started with a '49 Ford Chassis, then came the body components: front and rear fenders from a '51 Chevrolet, the hood from a '53 Studebaker, and a flat, rectangular grille formed from seamless tubing. A 331-cubic-inch '52 Cadillac V8 provided the horsepower. (Photo: Eric Rickman)

November 1957 Jack Simon's hybrid truck was a former '57 Ford four-door sedan. A '56 Ford pickup cab was grafted to the forward area of the sedan. Ford "500" quarter panels were employed as outer fender sections. Power was furnished by a "full house" '57 Pontiac V8. Jack's daughter, Jacquelyn, hoped to better the truck's 87-mph quarter-mile speed. (Photo: Gene Gauss)

March 1958 Rod and Custom magazine's "Truck for Tomorrow" was built according to reader suggestions. Four Buick turret top sections made the fenders and fins for the '50 Chevy. The rear of the contoured bed was welded solid, with no tailgate, and the rear window was cut much larger than stock to improve vision. The cab was sectioned above the fender lines. Because of extreme chassis lowness and wide-base tires, the fenderwells were heightened for wheel clearance. R & C editor Spence Murray headed up the popular project. (Photo: Bob D'Olivo)

September 1957 Henry McCormack, inspired by an HRM article on fiberglass construction, gleaned enough information to follow through with this excellent full-scale model of his own. He designed his dream car by completing a scaled drawing, clay model, plaster mold and fiberglass model. The car was only 49 inches high with a 106-inch wheelbase.

November 1957 Bill Bongers' '55 Chevy convert was lowered by chopping the front springs six inches and leveling the rear with six-inch blocks. The hood was nosed and smoothed, and the corners were rounded. The rear quarter panels were extended, '56 Lincoln taillights were fitted, and a '56 Packard bumper was molded into the sheetmetal. The stock '55 mill was swapped for the bigger '56 Corvette V8; it packed two four-throat carbs and high-rpm cam. A blue-on-white color scheme completed his project. (Photo: Peter Sukalac)

May 1958 Roy Dunn's Ford was a stylist's dream. A total of five inches was sectioned from the body, and full wheel cut-outs gave a sports car flavor. The side trim was from a '55 Lincoln. A ⅞-inch rod was used to form nice looking and functional nerf bars, and the grille was made from woven wire. (Photo: Eric Rickman)

July 1958 Leon Tetlow couldn't decide whether he wanted a custom or a rod, so he built both in one car. The club sedan was lowered four inches by modifying the frame and front and rear springs. The headlight shades were extended one inch, with '55 Cad rings bolted on. Taillights were from a '55 Packard; they were fitted to frenched openings in extended fenders. The mill had 335 cubes, and the heads were milled .030 and the passages polished. A Satin Gold lacquer paint job completed the look. (Photo: Bob Hardee)

April 1958 Dave Cunningham's '40 had a dropped axle, reversed spring eyes and kicked-up rear frame to lower the car. Clean lines were achieved by raising the fenders on the body. Reversed rims allowed ample tire clearance. The basic body modifications, such as channeling, cutting and mixing the color, were done by the owner. The '48 Merc engine had 296 cubic inches, a Potvin cam and the "usual port and relief job." Flywheel, headers, four carbs and an ignition completed the works. (Photo: Al Paloczy)

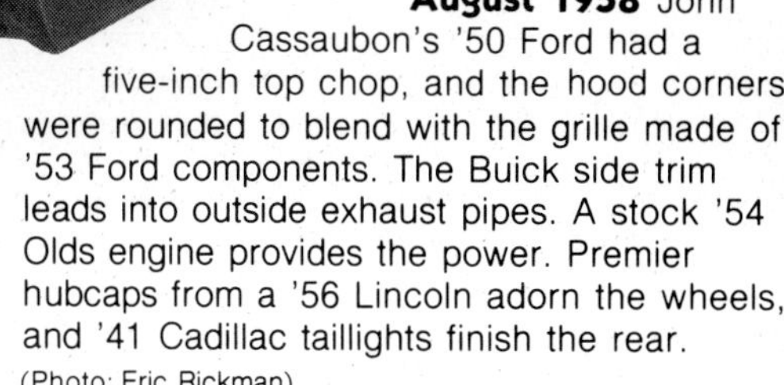

August 1958 Bill Carr's cover custom began as a '55 Chevy. He installed a ¾ race Corvette engine under the hood, then worked on the car's 48-inch height. This was achieved by a spindle kit at the front, de-arched springs, lowered blocks and "C'd" frame at the rear. The top was chopped 3½ inches, and the fins were extended 18 inches. The car was painted with 30 coats of Golden Honey Lacquer. (Photo: Al Paloczy)

August 1958 John Cassaubon's '50 Ford had a five-inch top chop, and the hood corners were rounded to blend with the grille made of '53 Ford components. The Buick side trim leads into outside exhaust pipes. A stock '54 Olds engine provides the power. Premier hubcaps from a '56 Lincoln adorn the wheels, and '41 Cadillac taillights finish the rear. (Photo: Eric Rickman)

October 1958 Ron Courtney's X-51 had just a hint of a rake. Full wheel cutouts blended well with the five-inch section job. Under the reshaped hood was a Chevy V8 for power. The fins were built from sheetmetal stock and were bonded to the quarter panels with fiberglass. The decklid was entirely rebuilt and the rear grille, made from tubing and reshaped sheetmetal, carried the theme from the front. A Fiesta Red paint job sealed the exterior. (Photo: Peter Sukalac)

February 1959 Charlie Fenza's convert was originally a two-door. Access to the Weiand and Isky equipped '53 Ford engine was gained by swinging up the entire front end, which was molded into one piece. Frenched headlights were raised with the fenders and hood. The Kaiser grille and a handmade lower pan were protected by bumpers once used as '56 DeSoto park lights. A Coral Flame and white lacquer paint job was divided by a combination of '52 Olds and '55 Pontiac side trims. Skirts were from a '55 Mercury. (Photo: Woody Higgins)

March 1959 Owner-builder Duane Aspengren sectioned his '49 coupe by trimming off four inches all around the bottom and channeling it an equal amount by raising the frame horns at each end. The filled, louvered hood overlooked a '57 Corvette grille housed in a '50 Merc shell. Unshaded headlights nestled within '52 Merc rings, which were both frenched and tunnelled. Duane matched performance to looks with a '56 Olds engine and Hydramatic. (Photo: Bob Hardee)

November 1959 Paul Plebauch's lowered '57 Ford Victoria was "neither overdone nor underplayed." The grille was a '54 DeSoto assembly, and the front bumper, chosen for its compatability with the grille, was from a '56 Ford. The front of the car got a whole new look with the addition of '56 Ford truck headlight rings. The Coral Mist paint was accented by darker scallops. (Photo: Allan Carter)

November 1959 Burt Hamrol's '50 Ford had modified windsplits to taper into Burt's own taillight design centered around '57 Chrysler lenses. Note the radiused wheelwells. (Photo: Dave Cunningham)

August 1959 The basis of Joe Brienza's Cad-powered '34 was the '34 cabriolet, which was chopped, channelled and sectioned for a height reduction of 19 inches. Part street rod, part custom, its radical molded rear body proved that Joe had the best of both worlds. (Photo: James Chasse)

December 1959 Studebaker didn't make a convertible in 1953, but Ray Moriarty took care of that little oversight with a power hacksaw and torch. The canted quad headlight openings, raised fender line and raised grille opening are all hand-formed. (Photo: Jim Chasse)

August 1960 Richard Gregg transformed his common '50 Ford Tudor into a full-custom pickup. The pickup bed rails were also exhaust pipes. The 1954 Pontiac taillight was hooded and frenched; body seams were filled all around. The burgundy paint was gold-striped to offset the truck's extreme lowness. The top was chopped five inches, and the body was sectioned the same amount. (Photo: George Barris)

November 1960 The "Impala Impulse" was a name coined by HRM to describe the customizer's urge to modify late-'50s model Chevys. Some of the modifications displayed in this article were bullet taillights, custom grilles, teardrop spotlights and scallop trim. Upholstered trunks attained a popular high, especially in show cars. Air scoops, dechroming, extended rear ends and recessed lights were also suggested. (Photos: George Barris & Colin Creitz)

January 1960 Fins on Eddie Johnson's '57 Plymouth reached six inches higher and four inches farther back than stock. The reworked torsion bar suspension dropped the tangerine-and-white custom seven inches. Special taillight trim and rolled tailpan gave further identity.

January 1961 Dean Jeffries gave HRM readers a lesson in panel painting. "No longer is a paint job merely a protective finish; now each custom becomes a mobile masterpiece with sharp detail, and overspray blends highlighting the original metalwork with bursts of color and accentuating curves."
(Photo: Eric Rickman)

December 1960 Ray Farhner started with a '32 frame and an incomplete roadster body; he built everything else. The top was chopped 3½ inches, and the body was channelled six inches. The white naugahyde interior was also installed on the running boards, inside the engine compartment and around the entire pickup bed. The tires were mounted on unusual '56 Chrysler spoke wheels.
(Photo: Grier Lowry)

July 1961 Vaun Hamlin, Jr. enhanced the original lines of his 1950 Ford coupe using several small refinements and a dramatic new grille improvised from '58 Pontiac metal. The stock front bumper was raised two inches and the hood shortened 1½ inches to match the new front end. One of his sharpest tricks was to simulate a body section job by reversing the rocker panel below the door. The concave panel was then painted black and the body was lowered to further the shallow illusion through the middle of the body. The rear wheelwells were also enlarged.
(Photo: Bob Hegge)

September 1962 Darryl Starbird was the owner of the "Wildest Tow Car in Town," a 1961 Thunderbird. It was chosen for its combination of good performance and comfortable luxury as tow car for the "Predicta," one of Starbird's excellent show cars. Modifications included a nosed hood and custom front end, which balanced the unusual taillight treatment. The headlight group was tunneled under a six-inch hood extension, rolled front and rear, and horizontal tubing adds to the visual width. (Photo: Tom Cusick)

October 1962 Bob Crespo started with a 1940 Ford Opera Coupe, and ended up with a "sweepstakes custom." The front end featured canted quad headlights, one-half-inch plastic rod detailing in the centers, and a floating grille shell with expanded metal and plastic rods. The hood was triple-scooped, and the fully molded body was almost entirely hand-formed. Although the coupe was unchannelled and unchopped, it had an eight-inch section taken out of it that posed many problems, especially in building a well-fitting hood. The front main leaf spring had reversed eyes, and the rear springs were arched to give rake to the coupe. (Photo: Dave Cunningham)

November 1962 Originally a '34 Ford convertible, Don Vargo's masterpiece was striking in Candy Apple translucent with a white naugahyde padded top. The body was channelled eight inches, doors were sectioned seven inches to clear the exhaust area. The grille shell holding the quad headlights was handmade with an expanded metal grille. (Photo: Alexander Bros.)

May 1963 Paul Savelesky's '55 Chevy hardtop received a lot of attention for being an "automatic custom." It could be started remotely, and it restarted itself automatically. The car door opened with a special light, and as it did, the seat automatically tilted out. The doors locked at 20 mph, and inside, there were a TV, hi fi and tape recorder. Not only the deck lid, but the doors, hood, windows and seats all operated electronically. (Photo: Pete Sukalac)

October 1963 Norm Porter's "Fifty with a Flare" started as a Ford ragtop. His curvacious metalwork made it into a very graceful blend, with Lincoln lights and a tube grille. The distinctive front bumper was made from a split '55 Pontiac component connected by tubing. (Photo: Pete Sukalac)

November 1963 Buddy Parazoo's Vicky was originally restyled by the Alexander Brothers, noted artists in the field of styling and customizing. But after Buddy acquired the car, he added a few more refinements. A shortened '59 Chevy bumper and mesh grille completed the front end of "Groovy Victoria." It was finished in Wild Cherry enamel. The reworked chrome and retouched lines produced an impressive, although conservative flavor. The wheels were the "new" mag style with bolt-in centers. (Photo: Pete Sukalac)

March 1964 Bill Cushenberry created this 1963 Ford Astro using interesting "round-plus-rectangle" styling from front to rear. The stock taillights were tunneled into the fenders, and a single bar was suspended in the grille cavity. The headlight treatment used Lucas lamps set in deep circular openings. Rear wheel openings were cut out, and the car was finished in a fine pearl with shading at the edges. Seats were white naugahyde, offset by turquoise carpeting. (Photo: Eric Rickman)

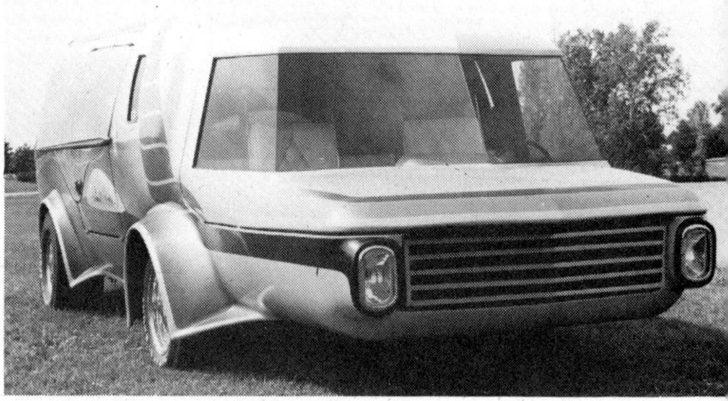

August 1967 Jess Leon's '32 Ford coupe was the culmination of four years of hard work. The sculptured body was painted in gold and pearl, and the upholstery, including inside the trunk, was pearl naugahyde and gold frieze. The hand-formed dash featured oak panels with Stewart-Warner gauges. The front axle was dropped 3½ inches and a '32 rear spring on a '40 Ford rearend lowered the rear of the car. The chromed nerf bars were formed from a ⅝-inch steel bar. "Unlike most customs, they are the real thing, designed to protect Leon's investment." (Photo: Bob Hegge)

November 1975 Darryl Starbird created this hand-formed "Vantasta," from what was once a 1974 Ford van. Shaped from 20-gauge hammer-welded sheetmetal and polished 5-inch polished flat aluminum stock, it was painted with candy purple and pearl white underbase with artwork by Crazy Paint. The stock 351 was covered by a Lucite see-through box.

August 1972 Jerry Pennington won the ICAS International Grand Champion and Custom Sweepstakes titles, as well as the HOT ROD Magazine Comet with his '69 Corvette, "Scorpion." "A true plastic fantastic," it was restyled by hand; spoilers, wheelwell flares and scoops were carefully molded into existing body lines. The red crushed velvet interior matched the Candy Red lacquer paint job. (Photo: Mike Brenner)

April 1974 The custom trend took a back seat to the more traditional street machines in the '70s, but the custom tradition was still evident in cars such as those owned by the Imperials of Los Angeles. Prominent elements of "lowrider" customs were smooth, clean lines, dropped suspensions and custom paint jobs. (Photo: Mike Brenner)

November 1977 Tom Philpot's '48 Merc mixed early styling with later-model add-ons. With a 289 Ford V8 engine, '70 Olds taillights, '58 Dodge hubcaps, VW sunroof, Philco stereo and "modern running gear," this car was a tribute to the early days of customizing. (Photo: Gray Baskerville)

Burning Up the

Hot rodders have always had a fascination with speed. The challenge, "How fast will it go?" has always been associated with modified cars. The removal of fenders, running boards and other unnecessary, heavy items in the quest of speed was an easy modification that helped form the look of the classic stripped-down hot rod. The search for even more speed led to engine swaps, usually in the form of the light and powerful Ford flathead V8.

Just making changes to the car wasn't enough. There had to be a way of testing it and challenging other car enthusiasts. Impromptu speed contests were the result, and the result of street racing was a public outcry against hot rodders. The dry lakes were a good place to test top speed, but they were a long way from town. Clearly, something else was needed.

The only straight, flat surface away from everyday traffic was the local airport. Early drag races were staged at airports until cities started building facilities just for drag racing. The NHRA Safety Safari was instrumental in helping many fledgling tracks get off to a good start.

As the facilities improved so did the cars and the speeds. (Elapsed times weren't even recorded in the early days.) Many slide rule jockeys claimed that it was impossible for any vehicle to attain speeds over 150 mph in the standing quarter mile, but were proven wrong. These same skeptics, it should be noted, keep raising their theoretical limits as hot rodders continue to prove them wrong.

Drag racing has become big business and a finely tuned science, so it is a real eye-opener to look back on the early cars when no single racer had the answer, and innovation was the password.

May 1970 "Wild Willie" Borsch was a Fuel Altered driver who always brought the fans to their feet with his hair-raising antics. With times in the low seven-second range at speeds over 200 mph, his short wheelbase altered was a handful, but that didn't stop Willie from calmly hazing the tires with only one hand on the steering wheel.

154
ORSCH

October 1952 Roger Hardcastle (center) and friends were among the 96 contestants that participated in the opening day event at the Redwood Timing Association's new dragstrip. Roger's five-window coupe used Evans speed equipment on the unusual fifth-mile track. (Photo: Ken Fuhrman)

December 1952 Bob Rounthwaite called his drag machine the "Thingie." The car ran at the opening meet held at the now-famous Pomona dragstrip. The engine in the "Thingie" was a 326-cubic-inch GMC. Bob was runner-up in the final race of the day when a blown spark plug eliminated the car's early lead. (Photo: W. G. Brown, Tom Medley)

December 1952 Dawson Hadley (left) in his '32 Ford coupe and Doug Hartelt in his track-nosed '34 Ford coupe were among the competitors at the opening race held at Pomona, California. The dragstrip, located at the Los Angeles County Fairgrounds, is still in use today. (Photo: W. G. Brown, Tom Medley)

July 1952 Louis Thompson displayed his competition coupe at the Century Topper's Auto Show in Modesto, California. The flathead engine used a Howard cam, Sharp heads and an Evans manifold. A radiator wasn't used. (Photo: Robert Moore)

October 1953 At the Pacific Southwest Championship Drags held at Paradise Mesa Airstrip in San Diego, California, Dode Martin's "A" dragster (top) beat the famed "Bean Bandits" car. The "Bandits" were awarded a trophy for the best e.t. of the day. (Photo: Eric Rickman)

February 1953 Back seat driving was the order of the day in this vintage Willys touring car. The driver sat behind a crude roll bar which barely covered a huge steering wheel. The engine was a Wayne head-equipped Chevy six. (Photo: W. G. Brown)

34 HOT ROD OF THE MONTH—

• "The West's Most Fabulous Dragster" is certainly a title befitting the Chrisman-Neumayer car. Not only is it a beautiful thing to see, it has power and performance to match its appearance. It is built around '28 Chev frame rails, Franklin front end and steering gear and a Model A rear end. A '39 Merc transmission is used in conjunction with a 10 inch Ford clutch and flywheel. The modified body was made many years ago. An eight gallon fuel tank is carried behind the body. Note the very sturdy roll bar. The engine is mounted directly to the frame at the front end and an engine plate is used at the back, mounted between the engine and gearbox. A special radiator ensures a cool engine when large dosages of fuel additives are used. The paint is metallic bronze lacquer, and the car is tastefully trimmed in red and chrome. A real sight to behold!

HOT ROD MAGAZINE

COPYRIGHT HOT ROD MAGAZINE

May 1953 A standard feature of the Hot Rod of the Month feature was a beautiful Rex Burnett cutaway drawing. In May the subject vehicle was Art Chrisman's 140-mph dragster. The car was actually owned by LeRoy Neumayer, but he turned it over to Chrisman when he entered the Army.

June 1953 One of the smallest dragsters ever featured in HOT ROD was the Crosley-powered rail of Nick Brajevich. Nick (left) was the inventor of Braje speed equipment for Crosley engines. The tiny terror was capable of 97 mph top speeds in the quarter mile. The little engine buzzed to 8200 rpm on straight methanol. (Photo: Bill Coleman)

October 1953 The HRM Club of the Month was the Drifters of Redondo, California. The club car was a '35 Ford with a 296-cubic-inch Merc engine. The fenders were bobbed and perforated to allow trapped air to escape from the fenders. The tires were 7.00 x 16 inch slicks. (Photo: Dean Moon)

January 1954 Eugene Robeck of San Jose, California, placed first in the "B" street roadster class at the Northern California Championship Drags. His time was 89.37 miles per hour. The flag starter was wearing his official pith helmet. (Photo: Bill Southworth)

January 1954 At the Northern California Championship Drags held at the Kingdon airstrip (Stockton), Dick Hubbard's Crosley-bodied "A" Modified coupe hit a top speed of 106.26. The Mercury-powered car also competed at the dry lakes and was featured on the cover of the January '53 issue. (Photo: Bill Southworth)

June 1954 Ed Losinski (foreground) was one of the top roadsters at the Pomona Valley Timing Association's So-Cal Spring Championships. The flathead engine used Sharp cylinder heads and intake manifold and a Harmon Collins magneto. The transmission was a '39 Ford unit with Zephyr gears. (Photo: Lester Nehamkin)

September 1954 Melvin Heath of Rush Springs, Oklahoma, brought his dragster to the NHRA-sanctioned Drag Safari at Caddo Mills, Texas. Heath's rail had a 351-cubic-inch Chrysler V8 and finished second in the final eliminations. (Photo: Eric Rickman)

October 1954 Definitely ahead of his time was Dick Katayanagi who raced his early Volkswagen at the Cal-Neva Timing Association Championship race. The event was run on a half-mile track. C. R. Johnston's Singer roadster easily defeated the VW. (Photo: Eric Rickman)

October 1954 Starting line action at the Paradise Mesa dragstrip pitted Harold Nicholson's Austin-bodied roadster against George Burkhart's '29 Ford roadster on deuce rails. Even though Burkhart's car was considerably heavier, he easily won the race. (Photo: Bob Greene)

November 1954 The first real drag meet ever held at the Linden airport in New Jersey was a fifth-mile event because the NHRA Safety Safari crew deemed the track unsafe for longer runs. The D.C. Dragons entered a very short dragster powered by a Mercury engine. The car had a roll bar and safety belt but the roll bar was below the driver's head. (Photo: Eric Rickman)

December 1952 In the days before exotic starting equipment, plain old manpower was the key ingredient for starting a balky car. This four-cylinder-powered Model A refused to cooperate with the crew at the opening race held at the Pomona, California, dragstrip. (Photo: Tom Medley)

December 1954 Jack Chrisman's sharp '29 Ford Tudor looked like a street machine but was actually a torrid dragstrip terror. The top was chopped 5½ inches, and the interior was gutted for racing. The radiator was filled in and covered with a wild pinstriping design by Von Dutch. (Photo: Eric Rickman)

July 1956 Dave Marquez set speed records and earned Best Appearing honors with his "B" roadster. He received help with the car from fellow members of the Motor Monarchs of Ventura, California. The car was powered by a mighty Ardun with Hilborn fuel injection.
(Photos: Bob D'Olivo)

December 1954 A lot of fine cars came out of the Chrisman garage in Compton, California, including Jack Chrisman's '29 Ford Tudor. Although it had the looks of a street machine, the car was a real dragstrip terror turning record time like 114.64 mph. The interior was gutted for racing and the flathead engine was set up to run on a mixture of 75 percent nitro and 25 percent alcohol.
(Photo: Eric Rickman)

December 1954 One of the best known racers ever to come out of San Pedro, California, was Frank ''Ike'' Iacono. His '33 Ford coupe was powered by a GMC six-cylinder engine topped by a Wayne head. The workmanship was excellent and times were spectacular—131.98 mph at Santa Ana. The paint job was a very bold combination of orange and black.
(Photo: Bob D'Olivo)

November 1954 A very early rear-engined dragster was the "Smokin' White Owl" of Ollie Morris (right). Harvey Malcolmson (left) sponsored the car and picked up the fuel tab, which was usually about $10 per weekend. At the time of the article, the car's best performance was 144.97 mph at Santa Ana dragstrip. (Photos: Bob D'Olivo)

January 1955 The Pacific Northwest Championship Drags were held at Scappoose, Oregon. Bob Bernstein's 275-cubic-inch Merc coupe faced off against Rodger Simonatti's 286-cubic-inch Merc-powered pickup with the pickup taking the win. Scorekeeping at the Pacific Northwest Drags was like that of most contemporary races—someone had to man a large chalkboard. In this case the recorder was Rex Delong. (Photos: Peter Sukalac)

December 1955 One of the more unusual and very successful dragsters at the first NHRA Nationals held in Great Bend, Kansas, was the "Bustlebomb" of Lloyd Scott. The dual-engined dragster featured an Olds engine in front and a Cad in back of the driver. The rear engine didn't have a flywheel, just a flange for a stub driveshaft leading to the ring and pinion. The car turned 151 mph at the first Nationals. (Photos: Bob D'Olivo, Eric Rickman)

August 1955 Ed Losinski's sleek, purple dragster was considered pretty spectacular in its day. The Chrysler-powered dragster used Hilborn injectors to put out 375 horsepower on an engine dyno and turn times as high as 143 mph in the quarter mile. The frame was made out of 3¼-inch chrom-moly tubing. The body was made out of aluminum.
(Photos: Eric Rickman)

January 1956 Considered a showpiece of the drag racing world, Ed and Roy Cortopassi's "Glass Slipper" was also very fast. The car turned a best of 130 mph in the quarter mile and reached 181 mph at Bonneville. The Sacramento, California, car used a full house Mercury flathead for power. The frame of the car was built from a pair of aluminum channel rails.
(Photos: Bill Chun and Bob D'Olivo)

January 1957 Ken Lindley of Anaheim, California, disproved the critics who said a dragster could never top 150 mph in the quarter mile by going 159.01 at the Nationals. The 331-cubic-inch Chrysler featured a front-mounted, crank-driven 6-71 GMC blower. (Photo: Eric Rickman)

February 1955 These photos of the famed "Bean Bandits" were published in '55 but actually shot in the summer of '51. The team was able to run in four classes, sedan or dragster with either one or two engines. This was before fiberglass bodies so it took quite a crew to remove the '31 Model A body.
(Photos: Don Cox)

January 1956 One of the most revered dragstrips of all time was the Lions Associated Drag Strip (LADS for short) in Long Beach, California. The best-known strip manager was Mickey Thompson (right) shown here with some of the trophies up for grabs.
(Photo: Jack Baldwin)

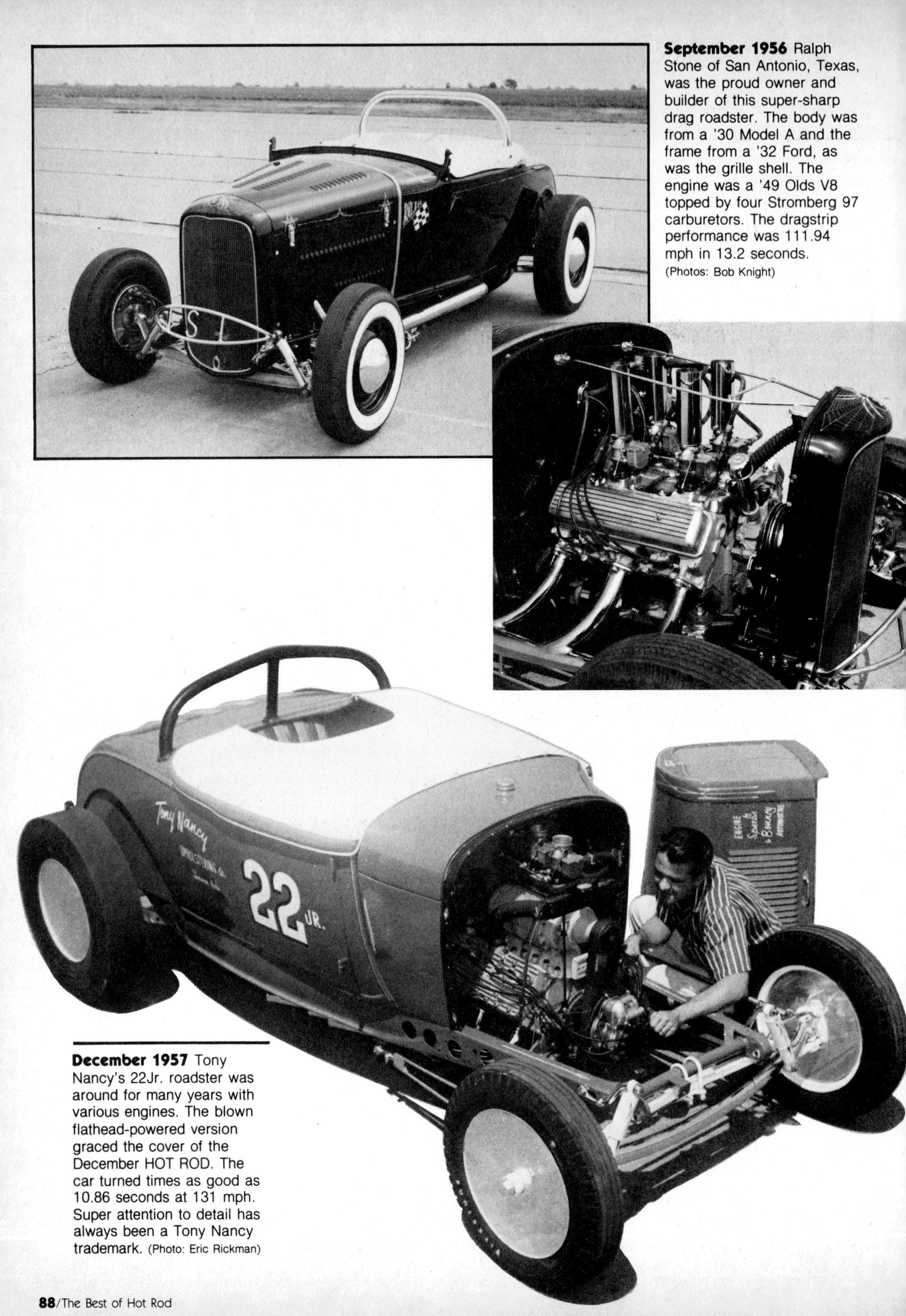

September 1956 Ralph Stone of San Antonio, Texas, was the proud owner and builder of this super-sharp drag roadster. The body was from a '30 Model A and the frame from a '32 Ford, as was the grille shell. The engine was a '49 Olds V8 topped by four Stromberg 97 carburetors. The dragstrip performance was 111.94 mph in 13.2 seconds. (Photos: Bob Knight)

December 1957 Tony Nancy's 22Jr. roadster was around for many years with various engines. The blown flathead-powered version graced the cover of the December HOT ROD. The car turned times as good as 10.86 seconds at 131 mph. Super attention to detail has always been a Tony Nancy trademark. (Photo: Eric Rickman)

April 1958 The HOT ROD Magazine Special driven by Cal Rice went after the International Acceleration Record at a special timing event held at March Air Force Base. Rice captured the Absolute World record for the Standing Kilometer, the International Class B record, the national Unlimited record, and the National Class B record. His top speed was 182.22 mph. (Photo: Eric Rickman)

November 1958 The Arfons brothers, Art and Walt, always had a flair for unusual machines, usually powered by big Allison aircraft engines and, later, jet engines. At the '58 Nationals, Art drove the "Green Monster" to the fastest speed of the meet, 156.25 mph. (Photo: Eric Rickman)

September 1958 Sam Parriott was a long-time participant at national drag races with a highly competitive and rare Kurtis sports car. The car was powered by a blown Cad which was tuned by "Big Bill" Edwards (kneeling next to the engine). Sam was later billed as the "World's Fastest Mayor" when he became mayor of the City of Industry, California.

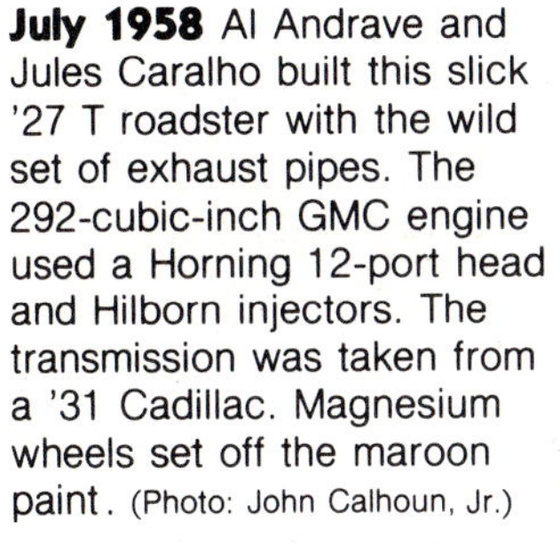

July 1958 Al Andrave and Jules Caralho built this slick '27 T roadster with the wild set of exhaust pipes. The 292-cubic-inch GMC engine used a Horning 12-port head and Hilborn injectors. The transmission was taken from a '31 Cadillac. Magnesium wheels set off the maroon paint. (Photo: John Calhoun, Jr.)

September 1957 Ted Cooper's Model A coupe really had readers scratching their heads over the wild set of exhaust pipes on the drag coupe. The pipes for the 351-inch Lincoln engine came up, over and under the set-back engine. The car was lowered through the use of a dropped axle, chopped top and channeled body.
(Photos: Bob Hardee)

November 1957 Charles Scott owned Scotty's Muffler Service in San Bernardino, California, and used his beautiful Ardun powered T to advertise his wares. Each cylinder had its own exhaust pipe that ran all the way to the rear of the car. The use of chrome plating was extensive.
(Photo: Eric Rickman)

November 1956 The cover car was Romeo Palamides' Oakland, California, dragster. Power came from a 375-horse DeSoto with Hilborn injectors. The car had a 115-inch wheel base and weighed 1550 pounds. The magnesium rear wheels were made by Palamides. (Photo: Bob D'Olivo)

November 1958 Smooth bodywork and quick times were the trademarks of Ed Norton and Armie Marion's Seattle, Washington, '25 T roadster. A 290-cubic-inch, Hilborn-injected '55 DeSoto engine powered the car to 11.08 e.t.'s at 128 mph. Front suspension was a Ford beam axle and the rear axle was from a '40 Ford with a Halibrand quick-change center section. (Photo: Allan Carter)

November 1958 Hank Bender (kneeling) and Ronnie Hier (in the car) proved that there was still a lot of horsepower left in the small-block Chevrolet engine. The 328-inch engine was featured on the cover and put out 430 horsepower on gas on Stuart Hilborn's dyno. The 1200-pound car terrorized big displacement dragsters for years on West Coast tracks. The team was from Santa Monica, California. (Photo: Eric Rickman)

March 1959 Chopped '34 Ford coupes were among the classics of dragracing machines. The front fenders were cut back on Frank Mosley's C/Altered coupe from Carlsbad, California. The engine was a 283-cubic-inch Chevy V8. (Photo: Eric Rickman)

February 1959 Don Garlits appeared many times in the magazine including this 1959 appearance. The '57 Chrysler engine was still carbureted. The dragster's frame was based on a pair of '30 Chevrolet rails. The best performance at the time of the story was 8.99 seconds at 167.55 mph. (Photo: Tex Smith)

October 1959 John Geraghty and John Crawford of Geraghty Automotive in Eagle Rock, California, built this gorgeous show-and-go machine in just two months at a cost of $3000. The Olds-powered beauty was known as "The Grasshopper" because of its brilliant green paint. Monogram Models later made a 1/24-scale plastic kit of the car. (Photos: Eric Rickman)

October 1959 The name of this car was the "Sidewinder" which accurately described the mounting of the blown Chrysler engine. A huge chain drive system transferred the power to the rear wheels. The car was owned and campaigned by the team of Chuck Jones, Joe Mailliard, Jack Chrisman and Wayne Reed. Performance was 9.03 seconds at 162.67 mph. (Photo: Don Nickles)

January 1960 One of the most bizarre-looking machines ever to run down the quarter mile was this '49 Plymouth coupe campaigned by the famed Ramchargers, who at the time were students at the Chrysler Engineering Institute. The engine was a 354-inch Chrysler with a homemade, early version of a tunnel ram manifold. The car looked silly, but ran strong.
(Photos: Eric Rickman)

November 1960 The beautiful red '29 roadster on '32 rails that graced the November cover belonged to the Cederquist brothers, Don and Dave, of North Hollywood, California. The grille shell was filled and decorated with the famous woodpecker symbol of Clay Smith Cams. The engine was a '54 Chrysler Hemi, rated at 400 horsepower.
(Photo: Curt Hamilton)

October 1960 Frank Maratta was the owner of a speed shop in Hartford, Connecticut, and of this clean '30 Ford Coupe. The tow truck was painted gold to match the race car. Moon discs were used on everything including the trailer. A '57 Chevy engine powered the car to a dragstrip clocking of 114 mph in 12.6 seconds.
(Photo: Jim Chase)

August 1960 Tommy Ivo (rear) campaigned Buick-powered dragsters for many years. Kent Fuller was the man responsible for the construction of Tommy's early cars. These two engines had a combined displacement of 934 inches and propelled the car to a speed of 173.7 mph in 8.69 seconds on gasoline.

December 1961 Ivo's outrageous four-engined Buick dragster was probably the most successful exhibition car of all time. The young man behind the wheel is Don Prudhomme who painted the car. Don also helped crew for Ivo before gaining tremendous fame as a dragster and funny car pilot. (Photos: Eric Rickman)

January 1961 1932 Ford Victorias are rather rare cars, especially in dragstrip form. The Vulcans car club of Wilmington, California, raced a sharp, purple Vicky that was powered by a Chrysler with an unusual blower setup. Updraft carburetors were used which accounted for the twin pipes coming through the cowl area. (Photos: Curt Hamilton)

January 1959 Frank "Ike" Iacono of San Pedro, California, was a big fan of GMC sixes with 12-port Wayne heads. He used a similar engine in his well-known '34 Ford coupe. The chassis for his dragster was built by Sonny Balcaen, and the body was formed by George Boskoff. The car ran 10.7 seconds at 130 mph on pump gas.
(Photo: Eric Rickman)

December 1960 Flag starters were still in use at the 1960 running of the U.S. National Drags. In 1960 the NHRA-sponsored event was held in Detroit, Michigan. (Photo: PPC Photographic)

May 1961 The cover photo (opposite) showed Holly Hedrich explaining the details of his rear-engined drag roadster to HOT ROD's Tex Smith. Holly and his partner Bob McClure based the sanitary machine in Altadena, California. The car's body was a streamlined '27 T. Best time was 9.62 seconds at 158 mph. Hedrich later served as publisher of HRM. (Photo: Eric Rickman)

March 1965 Bill and Bev Franklin's fiberglass '23 T was a real show and go piece. The 265-cubic-inch Chevy was run on nitro and alcohol to times of 10.97 seconds at 134 miles per hour. Halibrand mags were used on the Denver, Colorado, car. (Photo: Tony Spicola)

September 1961 Noted speed equipment manufacturer, Dean Moon, believed in running his products at the dragstrip. His bright yellow Mooneyes dragster was powered by a 300-cubic-inch Chevy with a crank-driven blower. The car ran American mag wheels, but the tow truck had the famous Moon discs. (Photos: Eric Rickman)

September 1962 One of the most feared dragster teams of the early Sixties was the team of Greer, Black and Prudhomme. The chassis was built by Kent Fuller and the body by Wayne Ewing. The rear of the body swept up to encase the parachute. The best performance at the time of the story was 8.09 seconds at 189.06 mph. (Photo: Don Francisco)

March 1963 Angelo Giampetroni of Gratiot Auto Supply in Detroit, Michigan, did more than just sell speed equipment. He used his wares in fine cars like this rear-engined '27 T which was powered by a 482-cubic-inch Buick engine. (Photo: Bob Hegge)

August 1962 Multi-engined dragsters were quite popular in the late Fifties and early Sixties. One of the really wild twin engine jobs belonged to Eddie Hill of Wichita Falls, Texas. The car used two 422-inch Pontiacs and four rear slicks for traction. The color was "Popsicle Purple" Metalflake and the car turned the quarter mile in 8.30 seconds at a speed of 190 mph. (Photo: Eric Rickman)

February 1965 It seemed like the Chrysler Corporation couldn't go wrong on the dragstrip in the mid-Sixties what with cars like the extremely popular Little Red Wagon. A 426-inch Hemi was installed in the bed of a Dodge A-100 compact pickup. The result was wild wheelstands and 10-second, 130-mph times. (Photo: Thomas E. Bedford)

July 1963 Back when innovation was the rule rather than the exception, there were a lot of unusual combinations on the nation's dragstrips. Jim Lakey of Piqua, Ohio, built a competition coupe using a Nash Metropolitan body and a '49 Olds engine with Algon fuel injection. The car received Rod & Custom Magazine's Best Appearing Car trophy at the '62 NHRA National Drags. (Photo: Eric Rickman)

February 1965 Even NASCAR superstar, Richard Petty, put in some time at the dragstrip. During a dispute over the use of the Hemi engine in 1965, Petty decided to build a Hemi-powered Barracuda and try straight line racing. This track at Piedmont, North Carolina, looks more like someone's driveway rather than a dragstrip. (Photo: Chrysler Photographic)

April 1965 By 1965, Don Garlits was running over 200 mph with ease. At the NHRA Winternationals in Pomona, California, he turned in the top run of the meet with a clocking of 206.88 mph. HOT ROD's editor, Bob Greene (right) presented Don with a trophy for top speed of the meet.
(Photo: Eric Rickman)

February 1965 One of the racers who really got the match race stockers going and helped evolve the cars into the wild funny cars was Dick Landy. Landy's '64 Dodge made extensive use of aluminum body panels including the doors, front fenders and hood. The wheelwells were opened up on the front edge and the car's wheelbase was moved forward for better weight transfer.
(Photo: Eric Rickman)

May 1965 The gasser wars of the Sixties were real crowd pleasers, and you couldn't ask for a better match-up than the two Willys of Stone, Woods, Cook (right) and "Big John" Mazmanian. In this match at Bakersfield, California, Doug "Cookie" Cook took the win at 9.90 seconds, 139 mph.
(Photo: Eric Rickman)

June 1965 Frank Pedregon claimed to have the world's fastest coupe with his wild Fiat Topolino-bodied dragster which ran 8.15, 195 mph at the United Drag Racers Association meet at Pomona. (Photo: Eric Rickman)

November 1965 There were wheelstands, and then there was the "Hurst Hemi Under Glass." Driver Bill Shrewsberry managed to get all four wheels off the ground at the '65 Indy Nationals. The Hemi in the back seat made for easy wheelies. The car was a certain crowd pleaser. (Photo: Eric Rickman)

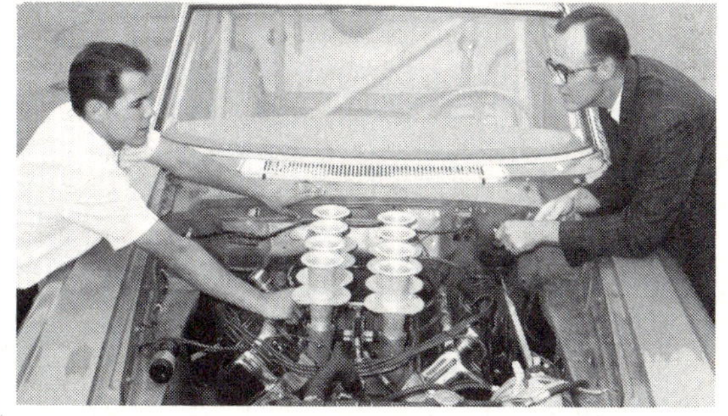

December 1965 Butch Leal (also known as the "California Flash") was a very popular A/FX (factory experimental) driver. His '65 Plymouth featured an altered wheelbase and injected Chrysler Hemi engine. Butch (left) took the time out from his busy schedule to explain the car to HOT ROD technical editor, Eric Dahlquist. (Photo: Eric Rickman)

June 1966 "Colt .45" was the name of Joe Davis' blown Chevy-powered Mustang funny car. The car's wheelbase was shortened one foot. Crew chief Gary Mallicoat (standing behind the car) maintained the 327-inch small-block. Dual batteries in the trunk aided the car's ability to do wild wheelies. (Photo: Eric Rickman)

October 1966 Doug Nash had a rather different funny car in the form of a fiberglass Ford Bronco. The car was quite light at 1450 pounds and used a Hilborn injected 289 engine and a C-4 automatic transmission. Times were in the 9.30-second range. (Photo: Eric Rickman)

July 1965 Lovely Joy Fernandez was more than just a pretty face around Norm Ries' wild T competition roadster; Joy was the top-billed member of the team of Fernandez, Ries and Hall. Joy helped maintain the flawless condition of the injected, Chevy-powered car from Cincinnati, Ohio. On initial trial runs the car posted a 9.84/148 best pass.
(Photo: Lynn Wineland)

April 1966 When Ford Motor Company got into drag racing they didn't hold back. Factory-sponsored cars like "Dyno Don" Nicholson's "Eliminator I" sported all the latest innovations. The engine was an SOHC 427 Ford that rested on a chassis built by Logghe Stamping in Detroit.
(Richard DeLonge and Don Rockhey of Ford Photographic)

November 1967 Few racing fans readily associate Don Prudhomme with Fords, but in the November '67 issue HOT ROD featured the Don Prudhomme-Lou Baney SOHC Ford 427 dragster. The engine was built by Ed Pink and the wildly striped paint job was done by Cerney. The AA/FD was capable of times like 6.88 at 223.82 mph. (Photo: Jim Kelly)

June 1969 The Ford Boss 429 was a rare engine, but Connie Kalitta had them in both his Top Fuel Dragster and "Bounty Hunter" funny car. Connie and the cars were featured on the June cover. The funny car used a B&M C-6 transmission while the digger relied on a Crowerglide. The Logghe Brothers built the chassis for both cars. (Photos: Norm Hanschu and Chuck Tupancy)

August 1969 The center spread featured one of the most successful Top Gas dragsters ever built, the fabulous "Freight Train." The twin, blown small-block Chevys powered the car to consistent 7-second blasts, always near the 200-mph mark. Bob Muravez was the "Freight Train's" engineer. (Photo: Eric Rickman)

April 1968 Bo Laws from Forest City, Florida, won the Street Eliminator title at the '68 NHRA Winternationals with his '67 Corvette. The car ran in D/Sports Production and was very consistent. (Photo: Eric Rickman)

April 1968 The cover photo featured this highly successful '56 Chevy sedan delivery stocker and its proud owner and driver, John Dianna. John joined the HOT ROD staff later that year and eventually became editor. Marv Ripes of A-1 Transmissions was responsible for the tough four-speed Hydramatic transmission in the car. The success of these transmissions later led to the outlawing of sedan deliveries from the stock classes. (Photo: Bud Lang)

January 1971 HOT ROD had a Corvette race car project in the form of Associate Editor John Dianna's '57 I/SA 'Vette. His partner, and noted automatic transmission builder, Marv Ripes drove the car to many important victories. The car was painted by Joe Andersen and covered in a how-to story. The powertrain consisted of a fuel-injected small-block backed by a Powerglide transmission. The combination was good for low 12-second times. In the photo, the young man nervously watching the car is John Dianna.

July 1970 A drag race wasn't complete without award presentations and trophy queens. At the 6th Annual S/S Nationals held at York US 30 Dragway, the team of Paul Candies and Leonard Hughes received the Cragar award from Ray Lavely (right). (Photo: John Dianna)

November 1978 One of the most popular and successful funny cars currently running is the Blue Max driven by Raymond Beadle. At the '78 Indy Nationals Beadle blasted into the five-second bracket with a 5.98-second run. He later won back-to-back NHRA/Winston World Championships in 1979 and 1980. (Photo: Tim Marshall)

April 1980 The best-known woman in drag racing and one of the sport's all-time winners is Shirley Muldowney, shown here on her way to the Top Fuel win at the NHRA Winternationals. She continued her winning ways that season to take the NHRA Winston World Championship for the second time (her first title was in 1977), a feat no one else has ever accomplished. (Photo: Tim Marshall)

March 1975 California was known for bracket racers that were both quick and good looking. John Rea's '55 Chevy Bel Air was chopped five inches and outfitted with a complete aluminum paneled interior. A 426 Hemi was set back 10 percent in the chassis and connected to a Torqueflite transmission. The car was capable of 10.30-second, 135-mph runs. (Photos: Bob McClurg)

November 1978 The caliber and performance of bracket racers in the late Seventies surpassed that of the Pro Stocks of just a few years earlier. Wayne Bidner's Camaro was one of the contenders at the Bracket Nationals. (Photo: Bruce Caldwell)

The Street Machines

In the early years, the emphasis was on hot rods and racing machines. Street machines began to edge into HOT ROD in the form of mildly customized late model Detroit cars. Even cars that were later considered street rods like '40 Chevy convertibles were considered street machines at the time. They were too new and not suited to the popular fenderless, cut-down hot rod look.

As the supply of good raw material for hot rods began to dwindle, many enthusiasts found themselves building street machines. By the time reproduction bodies were made available, the major auto manufacturers were building such great factory hot rods that the rush to street machines accelerated.

The trends in street machines were as changeable as any aspect of hot rodding. Fads changed from super low to mile-high suspensions to radical rakes to dragging rear bumpers. At times, engines were considered a mere necessity, yet when the street freak craze hit, monstrous motors were almost mandatory equipment.

Street machines offer great variety in the sport of hot rodding. Virtually everyone drives some form of daily transportation which could be transformed into a street machine. Wild or mild, street machines can be enjoyed by just about everyone.

June 1971 Corvettes always made great street machines, even in stock form, so when a big-block engine with a blower was added to a '62 Corvette, it was a winning combination. Paris Fish of Los Angeles built this refugee from "Route 66." The rear wheelwells were radiused, but that still wasn't enough to contain the 12.00x16 M&H slicks on Halibrand mags.
(Photo: Mike Brenner)

March 1952 Bill Grader of Seattle, Washington, chose a relatively rare '36 Ford roadster for the basis of his semi-custom street machine. Modifications included fender skirts, a recessed license plate, '37 DeSoto bumpers, '39 Ford taillights, a partially filled grille, electrically operated doors and a large headrest above the seat cushion. (Photo: Carl Payne)

February 1953 Among the celebrities present at the second annual Hamilton High School Hot Rod Jamboree in Los Angeles was noted customizer, George Barris (right). George presented the Custom Class trophy to Ken Lewis for his chopped, Carson-topped '41 Mercury convertible. (Photo: Eric Rickman)

September 1952 James Gaylord (of the hair care company of the same name) combined a '51 Ford with a Chrysler Hemi V8, and the result was billed as the ''Chicago Fordsler.'' The custom body touches included a shaved hood and a Canadian Metero grille. The car was painted golden bronze. Gaylord later built a very radical limited production car. (Photos: James Gaylord, Tom Medley)

October 1953 Chevys made good street customs as Jim Richardson proved with his sleek '41. The top was chopped 2½ inches, and the stock top was replaced by a padded Carson top. The interior was upholstered in pleated, maroon plastic with white piping.
(Photo: Eric Rickman)

April 1953 The Glaspar roadster was one of the first and best-constructed fiberglass kit cars. Don Cook built his Glaspar on a '39 Chevy chassis and installed a 302-cubic-inch GMC engine. The 180-hp engine was strong enough for a top speed of 98 mph at the Santa Ana dragstrip. The car took six months and $4000 to build. (Photos: Eric Rickman, W. G. Brown)

April 1954 One of Detroit's leading lady sports car club drivers, Marge Burrell posed proudly with S. E. Knudson's '52 Oldsmobile convertible. Modifications included a chopped top and a sectioned body. The engine was from a '52 Cadillac with 8:1 high-compression pistons. The wheels were Borrani wire wheels.
(Photo: Frank Burrell)

September 1954 Car clubs were big in the Fifties as illustrated by this group shot of the San Gabriel Valley Hi Domers. The California-based club consisted mostly of Fords and Chevys with a couple Mercs, a Pontiac and an Olds. (Photo: Tom's Photo Finishing)

April 1954 Teenagers Thom Charter and Ec Mackin drove Thom's custom Chevy convertible from Portland, Oregon, to Hollywood, California, to see the sights. Movie map mogul, Gilbert Morgan supplied them with the mandatory map of the stars' homes. After returning to Portland, the boys formed a car club, which had an immediate surge of activity, including two waxing parties and a gypsy tour. (Photo: Bill Southworth)

January 1954 College senior, Don Wells of Whittier, California, had to be a big man on campus with his sharp Chevy custom convertible. Modifications included removing most of the trim items and installing a one-piece Olds windshield. The car had a 110-volt A.C. converter to operate a PA system with microphone, record player and radio.

February 1956 Movie star Clark Gable was a hot rod and motorcycle enthusiast and had a McCulloch supercharger installed on his '55 Thunderbird. Shown with Gable is John Thompson of Paxton Products, McCulloch's distributors.

July 1957 Jim Gregg's '52 Chevy was modified by Wilhelm's Customs with items like a narrowed '56 Chrysler grille and Kaiser headlights. The side scoops were functional and the nerf bar was designed to surround the license plate. Altered A-frames lowered the front end. Jim belonged to the San Jose Gear Jammers car club. (Photo: Eric Rickman)

April 1956 '40 Ford coupes made great street machines and Whitie Tower's lemon yellow coupe was one of the best. Under the hood was a real eyeful—a 334-cubic-inch Merc with four Stromberg 97 carburetors mounted on a Weiand manifold. The wild interior was done in white and yellow naugahyde.

September 1957 One of the most famous pickups ever built was George Barris' '56 Chevy "shop truck." Most of the metal work was done by the very talented Sam Barris. The truck carried more gadgets than most homes including a TV, record player, telephone and a complete set of barber tools. The pinstriping was done by Dean Jeffries at the Barris shop in Lynwood, California.

November 1958 Jon Deston's '55 Ford from El Centro, California, was an all-around street machine. Besides the dechroming, the lowered stance and wild scallops, the car featured a '57 T-bird engine with three 97's. The paint was orange with green scallops. The Moon discs were louvered and pinstriped.
(Photo: Bob Hardee)

May 1958 Joe Zupan's '56 Ford F-100 pickup and model, Carole Austin, were featured on the cover of the May issue. Joe brought his truck from Pueblo, Colorado, to Barris Kustoms in Lynwood, California, to have the truck modified and painted. The paint job featured wild scallops done in Lime Gold and Organic Green. The spinner bars on the Olds Fiesta hubcaps were gold plated.
(Photo: Peter Jay)

November 1958 Dave Rolin's '57 Ford had a very distinctive look, to say the least. Headlights from a '57 Chrysler were tunneled under peaked fenders. The combination bumper and grille is from a '57 DeSoto. The front A-frames were kicked up for an eight-inch drop on the Sacramento, California-based cruiser.
(Photo: Dick Katayanagi)

December 1958 Longtime Petersen writer, cartoonist, editor, publisher, and now, Senior Vice President, Dick Day, has always had an interest in wild street machines. Dick's '55 T-bird featured a custom tube grille, nerf bars, chrome reversed wheels and conservative scallops.

December 1958 Customizer Joe Bailon is best known for his Candy Apple Red paint, but he also built many fine cars like this T-bird which belonged to Joe Castro. The front end was fully recontoured with tube grille bars, quad lights and straight nerf bars. The wheelwells were radiused and the chrome reverse wheels had bullet hubcaps.

June 1958 The new '58 Impala was an immediate hit with customizers and street machine owners so it was a natural that Bill Morse's mildly restyled Impala would make a good cover car, especially when Mamie Van Doren was added to the photo. At the time, Mamie was working on the film, "High School Confidential." (Photo: Jim Manatt, MGM Studios)

April 1959 "Miss Wichita" Ann Anderson was featured on the cover with Jack Lindsey's '55 Plymouth. The Wichita, Kansas, car was restyled by Darryl Starbird's Star Kustom Shop. '56 DeSoto side trim was used, the door handles were removed, the front fenders were extended and tunneled, and the car was lowered six inches. (Photo: Darryl Starbird and Tom Cusick)

MARFAK LUBRICATIC

The Best of HOT ROD

JANUARY 1948

MAY 1952

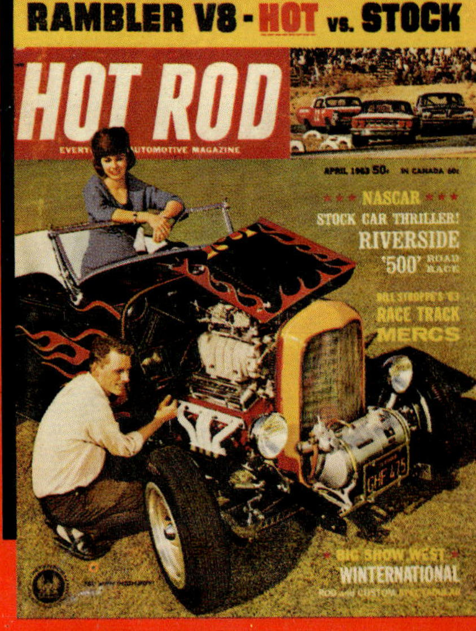

APRIL 1963

JULY 1978

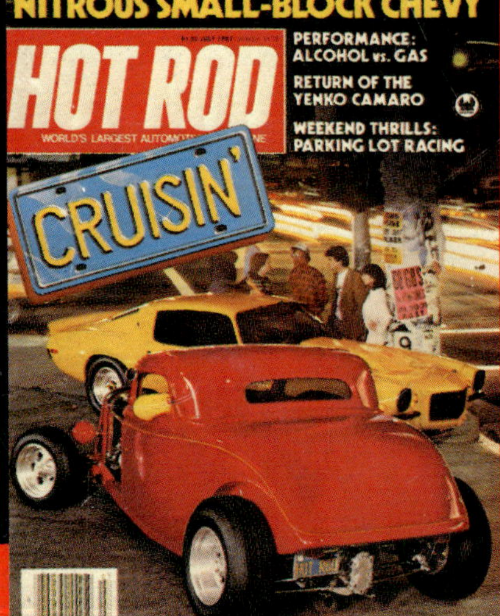

JULY 1981

HOT ROD Magazine is now in its fifth decade as the world's leading automotive publication. During those years, HOT ROD has covered every form of automotive endeavor. The competition scene has included circle track racing, all-out timed runs on the dry lakes, drag racing from its crude beginnings to the ultra-sophisticated cars of today, and the great tradition of the Indy 500. Modified street vehicles have been a big part of the HOT ROD editorial makeup, including classic street rods, cool customs and radical performance-oriented street machines. While reporting on engineering innovations, whether conceived by big manufacturers or backyard mechanics, HOT ROD has always made room for the unusual and bizarre in automotive expression. The Best of HOT ROD contains hundreds of superb photographs covering all these subjects and more.

ISBN 0-8227-5060-0